BOTANY.

ARTHUR, BARNES AND COULTER'S HANDBOOK OF PLANT DISSECTION. xi + 256 pp. 12mo. $1.20.

BESSEY'S BOTANY. (*American Science Series.*)
    Advanced Course. x + 611 pp. 8vo. $2.20.
    Briefer Course. xlii + 292 pp. 12mo. $1.08.

CROZIER'S DICTIONARY OF BOTANICAL TERMS. 8vo. $3.00.

HACKEL'S TRUE GRASSES. v + 228 pp. 8vo. $1.35.

KERNER & OLIVER'S NATURAL HISTORY OF PLANTS. With about 1000 illustrations and 16 plates in colors. Quarto. 2 vols.
    Vol. I. Bound in two parts. (*Ready.*) $7.50.
    Vol. II. (*In press.*)

MACDOUGAL'S EXPERIMENTAL PLANT PHYSIOLOGY. vi + 88 pp. 8vo.

UNDERWOOD'S OUR NATIVE FERNS. xii + 156 pp. 12mo. $1.00.

ZIMMERMANN'S BOTANICAL MICROTECHNIQUE. Translated by JAMES ELLIS HUMPHREY, S.D. xii + 296 pp. 8vo. $2.50.

Add 10 per cent. for postage if ordered sent by mail.

Publishers' Educational Catalogue free upon application.

HENRY HOLT & CO.,
29 WEST 23D STREET,   NEW YORK.

EXPERIMENTAL

PLANT PHYSIOLOGY

BY

D. T. MACDOUGAL

University of Minnesota

NEW YORK
HENRY HOLT AND COMPANY
1895

Copyright, 1895,
BY
HENRY HOLT & CO.

ROBERT DRUMMOND, ELECTROTYPER AND PRINTER, NEW YORK.

PREFACE.

THE appreciation shown toward the translation of Oels' *Pflanzenphysiologische Versuche* prepared by the present writer, together with the comments and suggestions from laboratories in which it has been used, has led to the preparation of this manual, which it is hoped will conform somewhat more nearly to the needs of American students. The general form of Oels' manual has been retained, and many cuts from the translation and a few paragraphs of the text have been repeated here without indication of their origin.

Only the more important and better established portions of the subject are treated, and these in the manner already in general use. With the rapid advance of investigation it is next to impossible that an elementary laboratory manual should include the latest results, especially when the essential points of many of them may yet be in controversy and need the critical treatment which is certainly not within the province of a work of this character. In the hands of an instructor in touch with current botanical thought, such deficiencies are easily supplied.

In the interests of precision, the term "assimilation" is here used exclusively to denote a general function of protoplasm, while the term "photosynthesis," which was introduced into the translation of Oels' manual (Preface and page 30), is adopted to signify the special process of forming carbohydrates from carbon dioxide and water in the presence of chlorophyll and sunlight.

The author is indebted to Mr. R. N. Day and Miss J. E. Tilden for the demonstration and drawing for Figure 32.

D. T. MACD.

MINNEAPOLIS, MINN., April 15, 1895.

CONTENTS.

	PAGE
Preface	iii
Introduction	1

I. ABSORPTION OF LIQUID NUTRIMENT.

Food of plants	3
Nutrient elements	3
Distilled water as a nutritive fluid	6
Influence of iron	6
Organs of absorption	7
Zone of root-hairs	9
Condition of nutrient substances in the soil	11
Nutrition of parasitic plants	10
Nutrition of saprophytic plants	11
Physical aspects of plants	12
Diffusion	12
Diffusion through epidermis	14
Power of selection of food-material	14
Turgor	16

II. MOVEMENTS OF WATER IN THE PLANT.

Root pressure	19
Transpiration	20
Evaporation of water from leaves	22
Wilting of excised shoots	23
Conditions of transpiration	24
Cause of wilting	26
Guttation	27
Attraction of soil for water	27
Uses of transpiration	28
Ascent of sap	28
Path of sap	33

III. ABSORPTION OF GASES.

Gases used by the plant	35
Diffusion of gases	36
Absorption of gases	36

	PAGE
Photosynthesis	37
Physical properties of chlorophyll	39
Division of the spectrum	40
Product of photosynthesis	41

IV. RESPIRATION AND OTHER FORMS OF METABOLISM.

Nature of metabolism	43
Respiration	43
Absorption of oxygen and excretion of carbon dioxide	44
Liberation of heat	45
Respiration essential to growth	46
Fermentation	47
Changes in color	49

V. IRRITABILITY.

Nature of irritability	50
Perceptive zone, motor zone	50
Geotropism	51
Perceptive and motor zones of roots	52
Amount of influence of gravity	54
Replacement of gravity	55
Heliotropism, thermotropism, etc.	57
Periodic movements	61
Hydrotropism	62
Contact movements	62
Circumnutation	66
Hygroscopic movements	66

VI. GROWTH.

Nature of growth	68
Grand period of growth	70
Influence of light on growth	72
Influence of light on anatomy of leaves	73
Influence of light and gravity on the formation of organs	74
Influence of temperature on growth	75
Sources of heat	76
Relation of temperature to distribution	76
Freezing of plants	77
Relation of moisture to freezing	78
Loss of heat	80
Resting period	80
Correlation processes	81
Mechanical force exerted by growing organs	81
APPENDIX	84
INDEX	87

EXPERIMENTAL PLANT PHYSIOLOGY.

INTRODUCTION.

A PLANT is a living organism which carries on, more or less constantly, certain life-processes. The more important of these are absorption of food-material, photosynthesis, respiration, transpiration, secretion, and reproduction. The manner in which these processes are performed is largely determined by the influence of the external conditions of gravity, light, heat, moisture, air, climate, etc.

In order to obtain an insight into plant life it is necessary to consider the nature, purpose, and mutual interaction of the life-processes involved and to analyze the influence exerted upon them by environment.

The course of experiments detailed in this manual deals only with the more salient features of plant physiology, and is illustrative rather than quantitative. In some instances, however, the simple treatment given may with proper application yield exact results. The physical and chemical apparatus possessed by every college or high school will be found sufficient to carry out the work.

Good plant-material is absolutely essential to the profitable performance of the experiments; and unless a greenhouse is at hand, the course should be pursued at a time when plants may be grown in the open air.

The following books will be found useful for reference:

DARWIN, F. Practical Physiology of Plants. 1894.
GOODALE. Physiological Botany. 1884.
KERNER and OLIVER. Natural History of Plants. 1894.
SACHS. Physiology of Plants. 1886.
SPALDING. Introduction to Botany. 1894.
VINES. Physiology of Plants. 1886.
VINES. Text-book of Botany. 1894–5.

METHODS OF EXPERIMENTATION.

THE entire course of an experiment should be described in detail in the student's note-book, with reference to the following points:

1. Object of experiment.
2. Apparatus and plant-material employed: condition and development of the latter. Full drawings.
3. Date of experiment and successive observations—day and hour.
4. Temperature, moisture, and sunshine.
5. Results of experiment.

CHAPTER I.

ABSORPTION OF LIQUID NUTRIMENT.

1. Food of Plants.—Green plants generally derive their food from simple chemical compounds in the soil, air, and water. Of these compounds the simple mineral salts are obtained from the soil. Animal manures and decaying vegetable matter do not serve directly as food-material. Elements and simple compounds liberated by the decomposition of these substances may be used in building up the plant. The chief value of such "organic" matter to the plant lies in the fact that it preserves the porous condition of the soil, thus allowing access of air to the roots and retaining water containing nutrient salts in solution. The decaying "organic" material in the soil also furnishes the proper conditions of growth for the soil Bacteria, whose activity is a necessary factor in the development of many of the higher plants. An admixture of "organic" material with the mineral elements of the soil is also a means of equalizing the temperature.

Aquatic plants in general use the same food-material as land plants. The water in which they grow is in contact with the soil and contains all of its soluble salts in solution.

Remark.—It is to be noted that the species comprised in the parasitic, saprophytic, and insectivorous plants, many of which are furnished with chlorophyll, are able to use directly complex substances derived from plants or animals, and do not depend entirely, or at all, on the simple compounds in the soil.

2. Nutrient Elements.—In order to determine what elements enter into the food of plants, and the office of each substance, plants may be grown in solutions of different composition. It has been found by numerous experiments and analyses that only potassium, calcium, magnesium, sulphur, phosphorus, iron, and sometimes silica and chlorine are obtained from the soil alone. Of the remaining substances necessary for the plant, hydrogen is obtained from both water and the soil compounds, oxygen from both the soil and the air, nitrogen usually from the soil, but in some instances also from the air, and carbon almost entirely from the air, except in the case of the plants which derive it from complex compounds. (§§ 1, 8, and 9.)

EXPERIMENT 1.

WATER CULTURES.

Fill a large bottle with distilled water, and to each liter of water add   or

1.	gram	potassium nitrate	1. gram	calcium nitrate
0.5	"	sodium chloride	0.25 "	potassium chloride
0.5	"	calcium sulphate	0.25 "	magnesium sulphate
0.5	"	magnesium sulphate	0.25 "	acid potassium phosphate
0.5	"	calcium phosphate		

Warm gently for an hour and keep in a dark place.

FIG. 1.

Hansen's germinator. *A*, glass vessel filled with water and covered with tightly-stretched bobbinet. *B*, bell-jar fitted with moist filter-paper. (Oels.)

ABSORPTION OF LIQUID NUTRIMENT.

Place seeds of Wheat, Corn, Bean, Pea, or Buckwheat in folds of moist cloth, in a pan of moist sawdust, or in a Hansen germinator (Fig. 1), until the radicles are a centimeter in length.

Fill a glass jar or cylinder of a capacity of 1 to 2 liters with the solution described above. The bottle should be well shaken before this is done. Add a few drops of iron chloride to the solution in the jar. Cut a hole 1 cm. in diameter through the center of a large cork and fit in the top of the cylinder or jar as shown in Fig. 2.

FIG. 2.

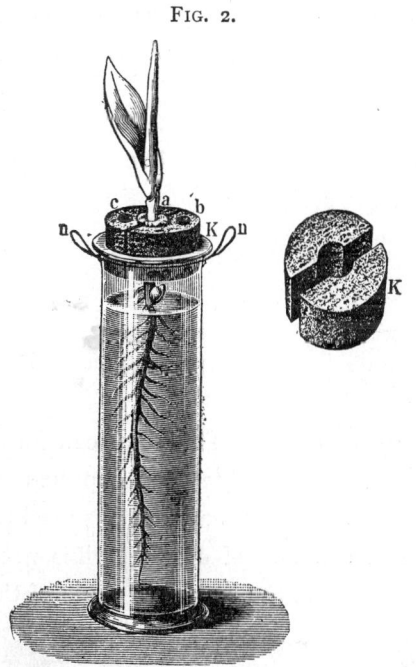

Culture cylinder with seedling of Corn in position. *K*, cork. (Hansen.)

Make a vertical cut from the outer edge of the cork to the central opening. Fasten a seedling obtained as above in the aperture by means of cotton-wool or asbestos fiber, in such position that the root only is immersed. Set in a sunny place. Renew the solution once every week.

Allow the plant to grow 3 or 4 weeks and compare with others of the same age grown in the soil.

Remark 1.—To exclude light from the roots and prevent the growth of Algæ in the culture cylinder it should be fitted with a jacket of pasteboard or blackened paper. This may be still more effectively accomplished if the jar is sunk its entire depth in the soil of a large flower-pot or box. If the soil is watered occasionally, a temperature more nearly suitable for the roots will be obtained.

Remark 2.—Calcium phosphate is only slightly soluble in water, and in consequence it forms a sediment on the bottom of the jar which decreases as that in solution is used.

Remark 3.—The sodium chloride used in the second solution is not of direct use to the plant, but serves to keep the solution alkaline.

Remark 4.—Analysis and agricultural practice show that plants of different species and genera grown in the same soil contain the elements in different proportions; consequently a solution suitable for all plants cannot be made. Instead of the salts given above, others which contain the elements in soluble form may be used. The degree of concentration must be kept within the prescribed limit, however.

3. Distilled Water as a Nutritive Fluid.—A plant will grow for a time in distilled water, but when the food stored in the seed is consumed it perishes.

EXPERIMENT 2.

DISTILLED WATER AS A NUTRITIVE FLUID.

Grow two seedlings as nearly alike as possible, one in distilled water and the other in a nutritive solution as in Experiment 1. Note difference in 10 and 14 days.

4. The Influence of Iron.—The plant can form green coloring matter (*chlorophyll*) only when supplied with iron. If this is withheld, the plant dies after it has used the iron stored in the seed. The presence of chlorophyll is necessary for the formation of food from the carbon dioxide of the air. (§ 29.)

EXPERIMENT 3.

IRON-FREE NUTRITIVE SOLUTION.

Grow two plants in an iron-free nutritive solution. The first leaves are green and the later ones pale yellow (chlorotic).

EXPERIMENT 4.

ADDITION OF IRON TO A CHLOROTIC PLANT.

Pour a few drops of iron solution into the culture jar of a chlorotic plant obtained by Experiment 3. The leaves soon become green. With a small brush moisten portions of a leaf of another chlorotic plant with iron solution. The portions treated will become green in a short time, and this color will gradually extend over the plant.

5. Organs of Absorption.—In the lower plants of simple organization the absorption of nutriment is carried on by a greater part or all of the surface of the organism. In the higher plants the roots are the special organs of absorption, and nearly all of the liquid taken in by the plant is obtained through them. The marked branching shown in the root sys-

FIG. 3.

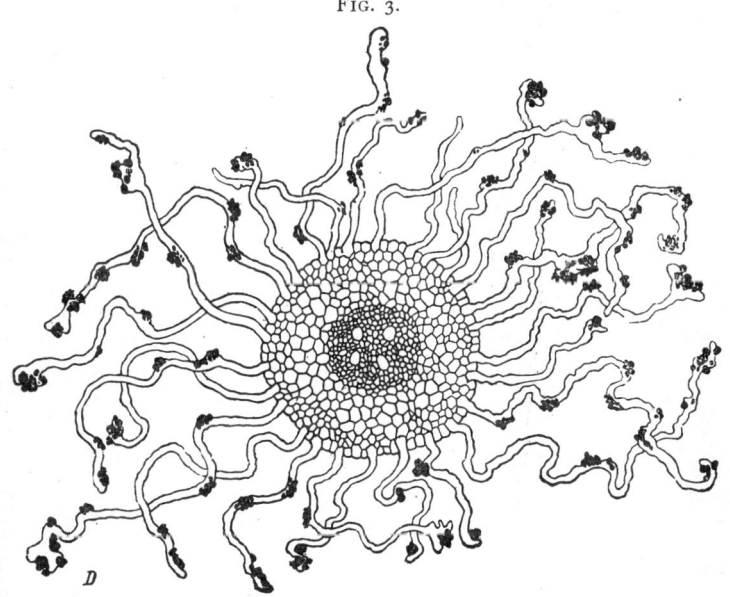

Cross-section of a root showing structure and arrangement of root-hairs. The latter are swollen in places, applying a broader surface to the soil-particles in contact with them. (Frank.)

tem of the higher plants not only increases the efficiency of the roots as organs for the fixing of the plant in the soil, but also magnifies the absorbing surface. Absorption is carried on through the outer walls of the peripheral cells which constitute the *epidermis*. In land plants the outer walls of these epidermal cells are developed into long tubelike extensions, the *root-hairs* (Fig. 3). The amount of surface extension obtained by root-hairs is very great, since these structures are .008 mm. to .14 mm. in diameter, and often attain a length of 3 mm., while from 10 to 400 may be formed on a square millimeter of surface.

The root-hairs are also an adaptation for obtaining water under the conditions in which it is found in the soil, where it occurs in the form of a minute layer on the surface of the soil-particles. The root-hairs are capable of bending around and penetrating between the particles in a manner which places their walls in contact with a large amount of this layer of water. In aquatic plants root-hairs are not needed, and are rarely formed, since the entire body of the root is in contact with the water. It will be seen that the land plants grown in water in the culture experiments developed very few root-hairs. On the other hand, plants grown in dry soil exhibit a very marked development of these structures. In this instance the amount of water around each soil-particle is very small, and the plant must reach a much larger number of them in order to obtain the needed supply.

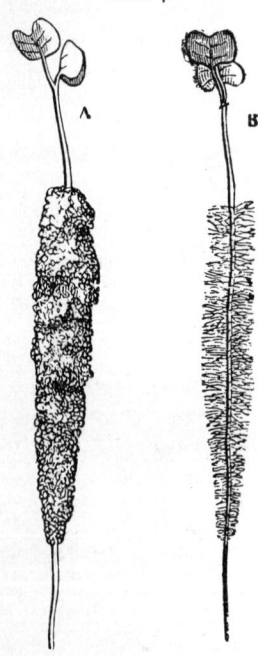

Fig. 4.

Seedlings of White Mustard. (Sachs.) *A*, with soil clinging to the roots; *B*, after removal of the mass of soil by washing.

EXPERIMENT 5.
ADHESION OF ROOT-HAIRS TO SOIL-PARTICLES.

Grow seedlings of Mustard, Pea, or Corn in sandy soil. When one week old take up and note amount of soil clinging to roots (Fig. 4). Free from the mass of the soil by washing. Examine with a magnification of 10 to 25 diameters and observe the remaining soil-particles attached to the irregular root-hairs.

EXPERIMENT 6.

STRUCTURE OF ROOT-HAIRS.

Cut a thin cross-section of the root of a seedling grown in the germinator and examine with a magnification of 50 to 100 diameters. Note the tubelike structure of the hairs, the thin irregular layer of protoplasm on the inner side of the wall and the large transparent central portion filled with sap. Examine the base of the hair and note its relation to the neighboring cells of the root (Fig. 3).

0. **Zone of Root-hairs.**—As the root extends in length by the growth of a portion near the tip, new root-hairs constantly arise in this region, while the older ones in the region farther away are constantly dying. The zone of root-hairs ordinarily begins 1 to 3 cm. back of the tip and extends backward along the root for a distance of about 5 to 10 cm. In this manner new root-hairs are continually brought into contact with fresh particles of soil.

EXPERIMENT 7.

MOVEMENT OF ZONE OF ROOT-HAIRS.

Place a germinated seed of Pea or Squash in a small funnel or thistle-tube in such manner that the root extends downward in the narrow outlet. Cover the seed with moist cotton and place a layer of moist filter-paper and a glass plate over the top of the funnel to prevent the seedling from becoming dried Set the funnel upright in a bottle containing a small amount of a solution of potassium hydrate. This solution will absorb the carbon dioxide gas given off by the seedling. By means of India ink mark on the glass tube the boundaries of the zone of root-hairs every day during a week. (Fig. 5.)

FIG. 5.

Apparatus to demonstrate progression of zone of root-hairs. (After Oels.) K, potassium hydrate solution; L, moist filter-paper.

7. Condition of Nutrient Substances.—Only liquid nutriment is taken up by the roots. The mineral substances of the soil are slowly dissolved by percolating water which contains small amounts of oxygen and carbon dioxide, as well as traces of nitric and sulphuric acids derived from the air. In some cases the walls of the root-hairs are saturated with an acid sap, which aids in the solution of the mineral salts.

EXPERIMENT 8.

ACIDITY OF ROOTS.

Place the roots of a seedling of Pea, Bean, or Corn grown in a germinator, on a sheet of blue litmus paper. The portion of the paper touched becomes red, indicating the presence of an acid.

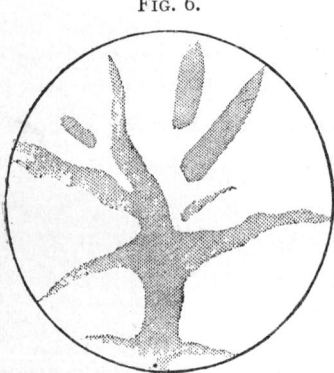

Fig. 6.

Marble plate corroded by roots. (Detmer.)

EXPERIMENT 9.

CORROSIVE ACTION OF ROOTS.

Fill a 5-inch pot half full of moist loam. On this lay a piece of marble whose upper surface is highly polished. Fill the pot with moist sand and imbed a germinated pea or bean near the surface. After the soil has been thoroughly penetrated by the roots (10 to 14 days) take out the marble plate, dry, and by reflected light note the rough lines etched on its upper surface by the acid of the roots.

8. Nutrition of Parasitic Plants.—Many plants of which Mistletoe and Cuscuta are examples do not develop a root system for the absorption of nutriment from the soil, but attach themselves to the bodies of other plants, from which they derive sap containing the necessary substances already prepared. Such plants are termed *parasites*. In many cases parasitic plants are entirely devoid of chlorophyll and depend altogether on the host-plant for their food. Cuscuta (Dodder),

Epiphegus (Beechdrops), and the microscopic Rusts and Smuts are examples of plants of this latter type. Mistletoe and many other parasitic plants are furnished with chlorophyll and are able to obtain a portion of their food-supply from the simple compounds.

EXPERIMENT 10.
NUTRITION OF CUSCUTA.

Examine plants of any ordinary species of Impatiens growing in wet or swampy ground, in September. On some of these plants may be found the yellowish cordlike twining stems of Cuscuta, bearing knotty masses of pale yellow or cream-colored flowers. With the plants still in position note that the Cuscuta has no soil-roots, but that it sends short *haustoria* or suckers into the stem of the Impatiens. With a sharp knife cut across the stem of the host-plant and determine the depth to which the haustoria have penetrated. The haustoria obtain sap from the host-plant by osmose, in a manner similar to the action of the root-hairs of land plants.

9. Nutrition of Saprophytic Plants. — Many plants derive all or a large proportion of their food-material from the products of the metabolism (see Chapter IV) of other organisms. Such plants are termed *saprophytes*. Examples of this type are afforded by Corallorhiza (Coral-root), Monotropa (Indian-pipe), Toadstools, Mushrooms, and many Bacteria.

EXPERIMENT 11.
NUTRITION OF TOADSTOOLS.

Note the growth of Toadstools and other Fungi on pieces of decaying wood in a damp forest. Tear apart the mass on which the plants are growing, and trace the long irregular absorbent organs ramifying in all directions through the mass.

EXPERIMENT 12.
NUTRITION OF MOULDS.

Place a fragment of saturated bread under a moist bell-jar for two days. A number of slender *hyphæ* of a Mould may be seen springing from the bread. Tear apart a small bit of the bread and examine with a magnification of 50 diameters. The absorbent *myceliæ* can be seen branching in all directions.

EXPERIMENT 13.
NUTRITION OF BACTERIA AND RELATED FORMS.

Make a solution of sugar in a cylinder and set in a warm room for three days. A film or scum will be formed on the surface of the liquid. Under a magnification of 600 diameters the scum will be found to consist of an immense number of globular, cylindrical, or spiral cells of Bacteria and other forms which obtain their food from sugar and complex substances formed by other plants. These organisms absorb food-material through their entire surface. The spores of such plants are found floating in the air and develop whenever they come in contact with food under proper conditions of temperature.

10. Physical Aspects of a Plant.—From a purely physical point of view the plant may be regarded as a cylindrical chamber whose walls are composed of membrane, and whose contents consists of a large number of stable and unstable compounds dissolved in water. At both ends of the cylinder the surface is magnified, at the lower end in the roots and root-hairs, and at the upper end in the leaves, to facilitate the diffusion of gases and liquids. The body of the cylinder, the stem, acts as a tubelike conductor between these surfaces. The outer layer of the stem is not easily permeable by fluids.

11. Diffusion.—By *diffusion* is understood any exchange which may take place between two fluids in contact either directly or through a membrane. This latter exchange is termed *osmose*. Diffusion takes place regardless of gravity until the fluids are alike. Not all fluids are capable of osmose, but only those which are imbibed by a membrane. The rapidity of diffusion varies with the mobility of the fluids. Stable compounds diffuse with more difficulty than water. Concentrated solutions of these compounds increase in volume, since they gain more water than they lose. They occur in the root-hairs, and in consequence a large quantity of water is taken up and forced into the root and upward in the stem.

EXPERIMENT 14.

OSMOTIC ACTION OF A SUGAR SOLUTION.

Cover the large end of a thistle-tube or small glass funnel with tightly-stretched membrane, such as parchment or bladder, which has been soaked for 15 minutes in water. Fill the large part of the tube or funnel with a solution of sugar 1 part and water 3 parts, and fasten upright by means of a large perforated stopper in a cylinder containing water, in such position that the two fluids are on a level. Note the height of the solution in the tube in 12 and 24 hours. A large amount of water has been drawn through the membrane into the sugar solution, while only a small portion of the latter has passed out into the cylinder, as can be ascertained by tasting. (Fig. 7.)

FIG. 7.

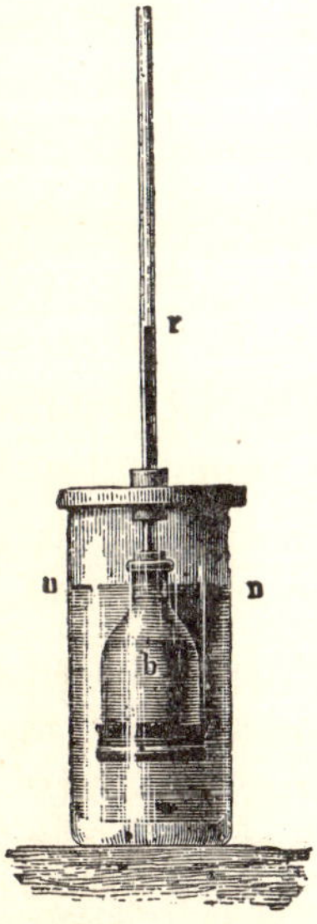

EXPERIMENT 15.

OSMOTIC ACTION OF A SOLUTION OF COPPER SULPHATE.

FIG. 8.

Cover one end of an ordinary lamp-chimney tightly with hog's bladder or parchment, fill with a solution of copper sulphate, and suspend in a vessel of

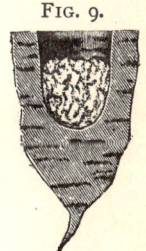

FIG. 9.

Osmometer. (Müller.) *b*, bulb of thistle-tube; *D*, level of liquid in cylinder; *r*, level of liquid in tube after a few hours' operation.

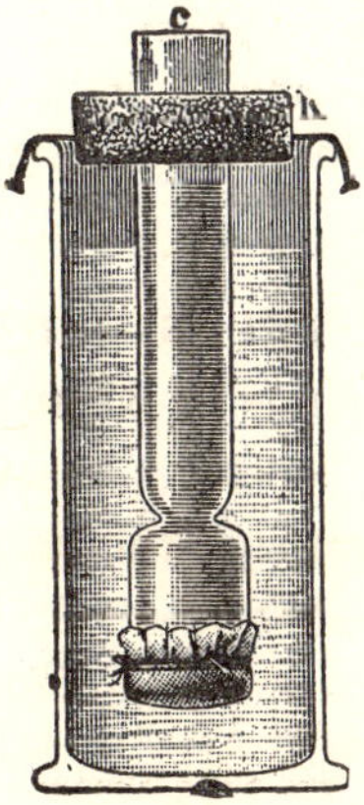

Osmometer. (Oels.) *K*, cork to hold lamp-chimney in place; *c*, open end of lamp-chimney.

Carrot hollowed out and filled with sugar. (Müller.)

distilled water. After a time the bluish color of the water in the vessel, and the copper-red coating formed on an iron nail placed in the vessel, denote that the copper-sulphate solution has passed through the membrane into the distilled water. (Fig. 8.)

EXPERIMENT 16.

OSMOSE IN PLANT-TISSUES.

Hollow out the central part of a large Carrot, making the walls of the cavity formed about .5 cm. in thickness. Fill the cavity with dry sugar. Twenty-four hours later the sugar will be dissolved in the sap which is drawn into the cavity, while the Carrot is dry and shrunken.

12. Diffusion through Epidermis of Aerial Organs.—Roots and root-hairs are pre-eminently organs of absorption, yet in some instances leaves and stems exercise this function. The leaf-like organs of Mosses and Liverworts are capable of absorbing water.

EXPERIMENT 17.

WATERPROOFING OF LEAVES.

Cut off a leaf of the Cabbage, Oak, Beech, or Iris and immerse in water. The surface takes on a silvery appearance, due to the thin layer of air adhering to it. An examination will show a heavy layer of cuticle or waxy substance on the outer side of the epidermis.

EXPERIMENT 18.

ABSORPTION OF WATER BY LEAVES.

Cut off a young branch of Coleus, Geranium, Tomato, Impatiens or other convenient plant and seal the end with wax or gum. Lay aside until slightly wilted. Immerse entirely in a vessel of water. Examine in two hours. If the leaves are capable of absorbing water, they will be restored to their original condition. It will be found that few plants can take water by means of the leaves. A moist atmosphere prevents loss of water, but does not form a source of supply for the plant.

13. Power of Selection of Food-material.—The root-hairs are immersed in a solution of mineral salts in the soil in a manner similar to the thistle-tube in Experiment 14. By the laws of diffusion all of these substances should be absorbed bo the root-hairs until an osmotic equilibrium is established, and generally such is the case. The amount of any one substance necessary to establish equilibrium is very small, and as soon as this amount is acquired, absorption of that substance ceases. When the plant withdraws any of the substances from the cell-sap solution to build up tissue, another quantity of that substance is absorbed from the soil. Thus different quantities of the various substances are absorbed. It is to be noted that all substances in the soil are not invariably extracted even in the minutest quantity by any one plant. The "rotation of crops" has its value for the farmer because different plants do not require the same soil-salts.

EXPERIMENT 19.

INCREASED DIFFUSION.

Close the lower end of two lamp-chimneys with bladder or

FIG. 10.

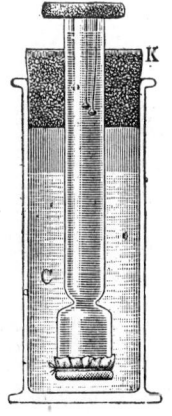
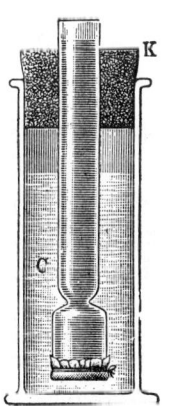

Apparatus to show selective diffusion. (After Oels.) *K K*, corks, loosely fitted ; *C C*, copper sulphate solution.

parchment and fill with distilled water. In the upper end of one cylinder place a stopper into which several iron nails have been driven. Make a saturated solution of copper sulphate and place exactly the same amounts in both cylinders. Now fasten the two chimneys in the cylinder as shown in Fig. 10. In 48 hours take out the chimneys and note that a bright deposit of copper has formed on the nails. Pour some granulated zinc into each cylinder. In 24 hours take out the undissolved zinc, and filter the solution in both jars, to obtain the copper precipitate. Allow the precipitate to remain on the filter-paper and dry. Weigh. It will be found that a much smaller quantity is obtained from the solution in the apparatus containing the nails. The action of the iron nails in withdrawing the copper from the solution inside the lamp-chimney, thus causing an additional amount to be taken up from the outside, will show the manner in which a plant exercises a "selective power" of absorption.

14. Turgor.—When a living cell, composed of protoplasm enclosing the cell-sap and surrounded by the wall, is placed in contact with water it absorbs the water in such quantity that the wall is stretched, while on the other hand the wall tends to contract by its own elasticity. Thus a cell-tension is set up which is denoted *turgor* (Fig. 11). The cells composing many of the tissues do not absorb water, while others take up large quantities and expand in consequence. If now a tissue which absorbs much water is attached to another which remains passive, a strain, or *tissue-tension*, will be set up. These tissue-tensions give rigidity to herbaceous plants. The wilting of plants is accompanied by loss of turgor, and consequent decrease of the tissue-tensions.

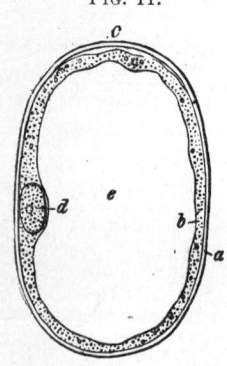

FIG. 11.

Diagram of cell. (Hartig.) *a, c*, wall; *b*, protoplasm; *d*, nucleus; *e*, cell sap.

EXPERIMENT 20.

TURGOR IN AN ARTIFICIAL CELL.

Cover one end of an open glass cylinder 10 cm. in length (a large tube will suffice) with membrane, fill with a sugar solution, close the other end in the same manner, and place in a vessel containing water. The contents of the cylinder increase in volume by absorption of water, and the membranes take a convex form in consequence of the increased pressure inside the cylinder. Place the cylinder in a vertical position and pierce the upper membrane with a needle. The liquid spurts upward from the pressure. (Fig. 12.)

FIG. 12.

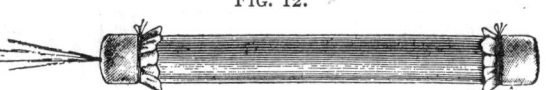

Artificial cell to illustrate force of turgor. (After Oels.)

EXPERIMENT 21.

IMBIBITION (OSMOSE) OF WATER BY SEEDS.

Ascertain the exact weight of 20 dried Peas, and place in a dish containing distilled water. In 24 hours the Peas will have greatly increased in size by the diffusion of water through the seed-coat. The substances stored in the seed have a strong attraction for water. Dry the seeds by rubbing with a cloth, and weigh. In some cases they will have taken up their own weight of water. The force of the osmose may be shown if a bottle of 25 cc. capacity is filled with the seeds and immersed in a vessel of water for 24 hours.

EXPERIMENT 22.

LONGITUDINAL TISSUE-TENSIONS.

Cut a slice a centimeter in thickness and 10 cm. in length

FIG. 13.

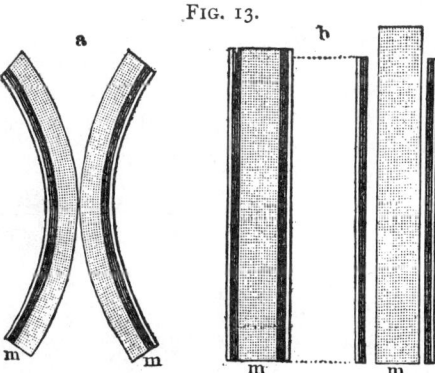

Longitudinal tissue-tensions. (Hansen.) *a*, outward curvature of a section due to excess of turgor of pith; *b*, length of separated tissues; *m m*, pith (parenchyma).

from the centre of a young branch of Elder (Sambucus) or stem of Rhubarb. Divide the slice in halves and note the outward curvature of the two parts. Describe the tensions which existed. Prepare another slice and separate the parenchyma (pith) from the wood and epidermis. The parenchyma expands and the other portions contract. (Fig. 13.)

EXPERIMENT 23.

TRANSVERSE TISSUE-TENSIONS.

Cut a ring of bark from a young twig of Willow or Poplar, and after a few minutes replace in its original position. It now does not extend entirely around the twig. When in that position it must have been in a stretched condition. (Fig. 14.)

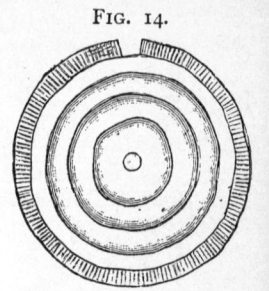

FIG. 14.

Transverse tissue-tensions. (Detmer.)

CHAPTER II.

MOVEMENTS OF WATER IN THE PLANT.

15. Root-pressure.—The roots, by reason of the osmotic activity of the substances which they contain, are constantly absorbing water. The amount taken up during the winter season when the soil is either frozen or at a very low temperature is very small. At the beginning of spring the storage products which were accumulated in the roots during the latter part of the previous season are changed into substances, such as sugar, dextrine, asparagin, etc., which are soluble and possess great osmotic activity. At this season the leaves are not yet formed, and only a limited amount of water is carried up and transpired by the plant; consequently the water taken in by the roots is slowly forced upward in a stream, almost filling the wood-cells in the lower part of the plant. The action of the roots is well illustrated by the osmometer described in Experiment 14. The pressure with which the water is forced upward by a Nettle will sustain a column of water 3 or 4 meters high. In the Grape the *root-pressure* is sufficient to sustain a column of water 10 meters in height. A yearly periodicity of root-pressure is noticed in trees and other perennial plants. In addition it can be demonstrated that daily variations due to temperature of soil and air and the humidity of the air occur. In the Grape the pressure is greatest in the forenoon, and decreases from 12 to 6 P.M. The root-pressure of the Sunflower reaches its maximum and begins to decrease at 10 A.M.

EXPERIMENT 24.

MEASUREMENT OF ROOT-PRESSURE.

Cut off the stem of an actively-growing plant of Dahlia, Geranium, Corn, Sunflower, or Grape a short distance above the ground, and fasten tightly to the stump in a perpendicular position a long glass tube by means of a short section of rubber tubing. Observe the varying height of the sap in the tube from day to day, noting the temperature and moisture of the air at the same time. (Fig. 15.)

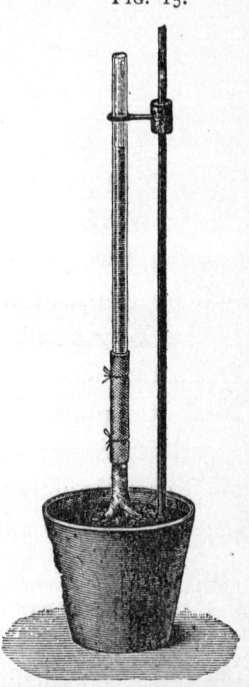

FIG. 15.

Apparatus for demonstration of root-pressure. (Detmer.)

16. Transpiration.—The water taken up by the roots finds its way upward through the stem toward the leaves, where a constant diffusion into the air takes place. The diffusion of water from the leaf or other organs of a plant into the air is designated *transpiration*. Transpiration takes place under the same physical laws as the evaporation of water from a moist membrane. Barometric pressure, light, temperature, humidity, and movements of the air are the most important conditions affecting the process. The amount of water actually given off varies also with certain metabolic processes (see Chapter IV.). A plant may be compared to a tube filled with water, with an expanded upper end closed by a membrane, while the lower end is immersed in water. By evaporation an upward-flowing stream is set in motion.

EXPERIMENT 25.

LIFTING-POWER OF THE EVAPORATION OF WATER FROM A MEMBRANE.

Fill a thistle-tube by placing it in a vessel of water, and while in that position cover the large end by a tightly-stretched membrane.

MOVEMENTS OF WATER IN THE PLANT.

Close the small end of the tube with the finger, lift from the water, and place in an upright position with the small end immersed in a dish of mercury. (Fig. 16.) Examine daily for two weeks or more. As the water evaporates the mercury slowly rises. It may be drawn to a height of 34 cm. if a good quality of ox-bladder is used.

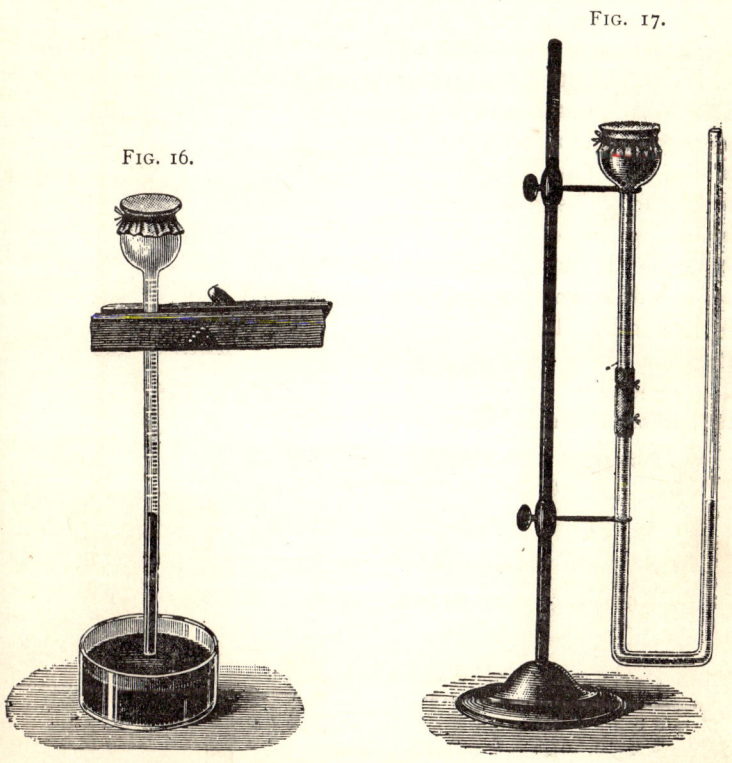

FIG. 16.

FIG. 17.

Apparatus to demonstrate lifting power of evaporation. (After Oels.)

This experiment may also be carried on in the following manner to determine the amount of water evaporated: Fit a thistle-tube with a membrane as above, and while still under water attach to it by means of a short section of rubber tubing a glass tube bent twice at right angles (Fig. 17). Completely fill the apparatus with water and fasten in an upright position. Place a drop of oil on the surface of the water in the open tube to prevent evaporation at this point. The amount of evaporation from the membrane will be shown directly by the fall of the level in the open tube.

17. Water Evaporates from Leaves as from a Membrane.

—The shoots and leaves of plants give off water in a manner similar to the action of the apparatus in the above experiments.

EXPERIMENT 26.

LIFTING-POWER OF TRANSPIRATION.

By means of a closely-fitting rubber stopper fasten a leafy shoot of a woody plant (Raspberry, Rosebush, etc.) in one end of a U tube filled with water and mercury. The mercury in the open arm of the tube soon begins to sink, indicating a loss of water from the leaves. (Fig. 18.) This fact may also be demonstrated by fitting the shoot to the upper end of a straight tube whose lower end is

Fig. 19.

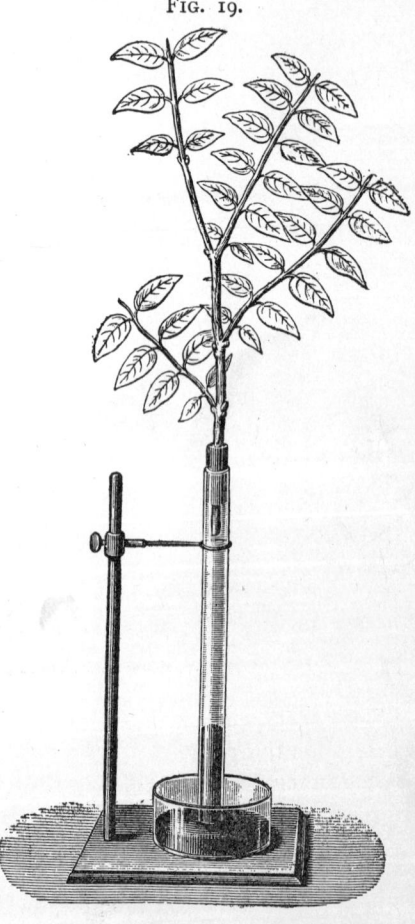

Lifting power of transpiration. (Detmer.)

Fig. 18.

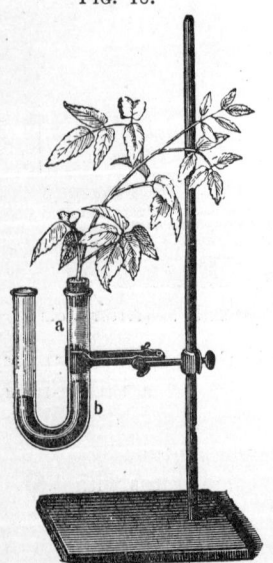

Lifting power of transpiration. (After Oels.) *a*, **water**; *b*, mercury.

immersed in mercury. (Fig. 19.) The lifting power of transpiration can be estimated from the height of the mercury column.

EXPERIMENT 27.

ESTIMATION OF THE TRANSPIRATION FROM A SINGLE LEAF.

Fasten a leaf with a round smooth petiole in one end of a U tube by means of a rubber stopper. Previously fill the U tube with water and fit in the other end a long capillary tube bent at right angles. Place the apparatus in such position that the leaf will be

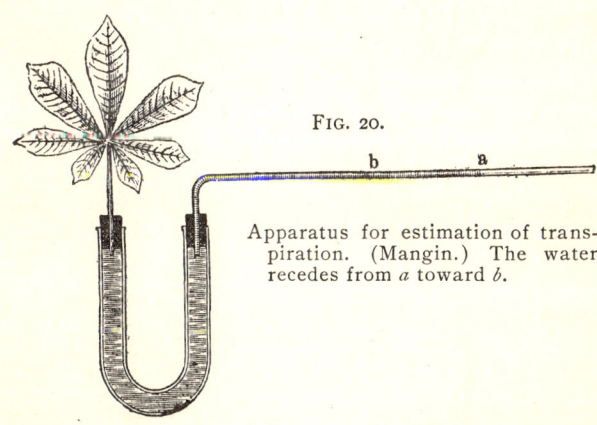

FIG. 20.

Apparatus for estimation of transpiration. (Mangin.) The water recedes from *a* toward *b*.

held upright, and the long arm of the small tube horizontal (Fig. 20). A small amount of transpiration from the leaf will cause the water in the small tube to recede horizontally. The amount and rate of transpiration may be easily computed.

18. Wilting of Excised Shoots in Water. — Herbaceous shoots when cut off and set in water generally wilt quickly, but if the shoot is cut under water it remains fresh a much longer time. Evidently the cause of the wilting when the stems are cut without this precaution is the penetration of the shoot by air. Perhaps in rapidly-growing plants escaping slime may seal up the ends of the vessels which conduct water.

EXPERIMENT 28.

WILTING OF SHOOTS EXCISED IN AND OUT OF WATER.

Bend a long shoot of a slightly woody plant (Symphytum, Rosebush, etc.), so that a portion of the stem is under the surface of the

24   EXPERIMENTAL PLANT PHYSIOLOGY.

water in a dish. Cut off the stem under water, and it will remain fresh several days if the cut end is kept immersed. At the same time cut off another shoot in the air, and after 10 minutes place the

FIG. 21.

Excision of a shoot under water. (After Oels.)

cut end in the vessel of water with the other shoot. Compare results daily. (Fig. 21.)

19. Conditions of Transpiration.—Plants transpire water constantly over their entire aerial surface, yet the stems and the

FIG. 22.

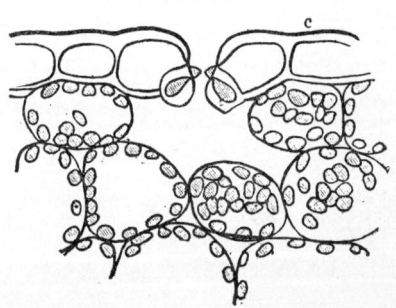

Stoma from under side of leaf of Iris florentina. *c*, cuticle. (Strasburger.)

greater part of the leaves are covered with an almost impervious layer of cuticle. Beside this, the devices exhibited by plants, especially those growing in the drier regions, by which transpiration may be lessened or controlled, are very numerous. One of the most effective is a covering of bristly hairs. Much the greater part of the water thrown off by the plant is transpired from the thin-walled cells in the interior of the leaf into the inter-

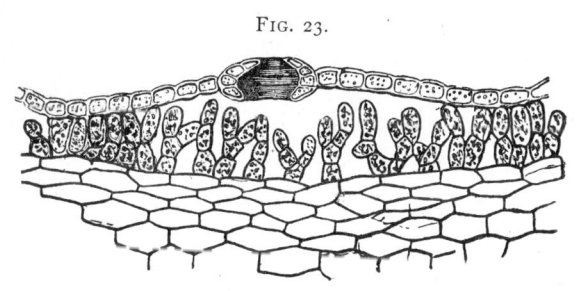

FIG. 23.

Air-chamber and opening of Marchantia polymorpha : magnified 300 times. (Kerner.)

cellular spaces which communicate with the open air through the stomata. The *stomata* (Figs. 22 and 23) are openings in the epidermis, which are controlled by *guard-cells*. When more water is transpired from the leaf than is furnished by the roots, the guard-cells become flaccid, and the walls are thickened in such manner that in this condition these cells change their form and close the openings of the stomata entirely. When the necessary water-supply is at hand the guard-cells are turgid and the stoma remains open. The action of the guard-cells is also influenced by light, wind, and other factors. Transpiration is increased by heat, light, dryness, high pressure, and movements of the air, and lessened by the opposite conditions.

EXPERIMENT 29.

INFLUENCE OF HUMIDITY ON THE AMOUNT OF TRANSPIRATION.

Place a well-leaved Begonia grown in a pot, on one pan of a druggist's balance. Cover the soil by means of two glass plates, or tie a piece of oiled cloth around the entire pot, to prevent evaporation.

By means of weights on the other pan equalize the balance. In an hour it will be noted that the end of the scale holding the plant has risen. Take weights from the other pan until the equipoise is restored. The amount of the weights taken off will represent water transpired by the plant. After balancing cover the plant by means of a bell-

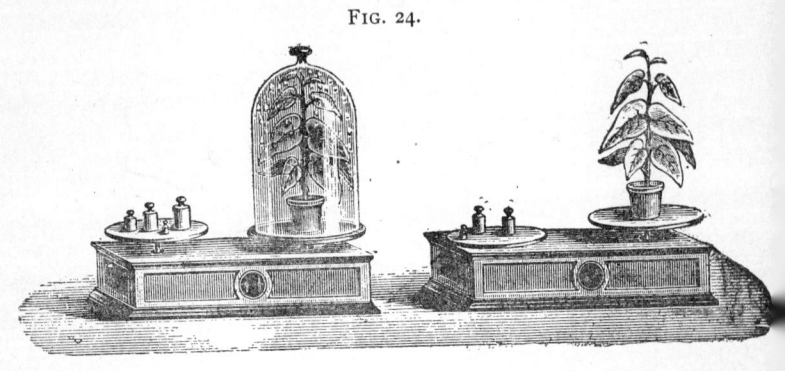

FIG. 24.

Estimation of amount of transpiration by weighing. (After Oels.)

jar. In an hour remove the bell-jar, quickly wipe from the pan the water which may have condensed and run down the sides of the bell-jar, and again take off weights to balance. The amount lost will be less than before. The air in the bell-jar soon becomes saturated with water and checks transpiration. (Fig. 24.)

EXPERIMENT 30.

INFLUENCE OF EPIDERMIS ON TRANSPIRATION.

Select two Apples and two Potatoes of equal size. Peel one of of each. Weigh and set aside for three hours. Again weigh. It will be seen that a waxy or corky epidermis retards transpiration very efficiently.

20. Wilting.—If the amount of water transpired exceeds that absorbed by the roots, wilting results. This may occur from the destruction of the root-hairs or from an insufficient supply of water in the soil. In the transplantation of trees the branches are trimmed in order that the transpiring surface may be reduced in proportion to the absorbing surface. The latter—in the root-hairs—is nearly all destroyed by transplatation. The *turgor* of a wilted plant may be restored either by watering the soil or checking transpiration.

EXPERIMENT 31.

RESTORATION OF A WILTED PLANT BY CHECKING TRANSPIRATION.

A plant if not too badly wilted will revive if placed under a bell-jar or if *transpiration* is checked by other means.

21. Guttation.—If the amount of water absorbed by the roots is in excess of that transpired by the leaves, it will exude through rifts in the epidermis, or the *water-pores*, in the form of drops. This process is termed *guttation*. It may be observed in plants at the end of a warm day. The air cools quickly, and its relative humidity is increased while the roots absorb the same amount of water from the soil, which retains its warmth for a longer time.

EXPERIMENT 32.

GUTTATION PRODUCED BY CHECKED TRANSPIRATION.

Cover a plant such as Corn, Wheat, or Pea with a bell-jar and place in sunlight. Note the drops of water on the leaves after an hour or two.

22. Attraction of Soil for Water.—Plants cannot either by the force of diffusion or of transpiration absorb all of the water in the soil. Absorption finally reaches a limit beyond which the capillary attraction of the soil-particles for water is stronger than the combined force of diffusion and evaporation in the plant.

EXPERIMENT 33.

AMOUNT OF WATER IN THE SOIL WHICH CANNOT BE ABSORBED.

Grow a plant (Bean) in a pot filled with rich garden soil. As soon as the primordial leaves have developed, place in a room exposed to direct sunlight, and allow it to remain without watering until it wilts. Now take a sample of a few grams of the soil which has been penetrated by the roots, and dry at 100° C. for an hour. Weigh. It is demonstrated that the soil contained a large percentage of water which the plant could not obtain to replace its evaporation.

23. Uses of Transpiration.—In the economy of the plant transpiration is of the greatest importance. Water and dissolved nutrient salts are carried to the leaves by the transpiration stream. The greater part of the water evaporates, and the remainder, with the salts, is formed into compounds useful to the plant. In the leaves the simple "power of selection" operates as in the roots, and only the salts they can use are carried to them in quantity. It is probable that transpiration serves other uses which are not yet clearly understood. The suggestion has been made that it equalizes changes of temperature in the plant.

24. Ascent of Sap.—The forces concerned in carrying water from the roots to the leaves are root-pressure, capillary action of the wood-cells, imbibition, diffusion, expansion and contraction of the air-bubbles in the wood-cells, transpiration, and osmotic action of the protoplasm of the wood-parenchyma cells.

In small herbaceous plants root-pressure is almost always present, and it acts with a force sufficient (see paragraph 15) to drive water to the leaves. In plants of this character the suction exerted by transpiration is also sufficient to carry water upward to the desired height. (See Experiment 26.) The other factors mentioned are of minor importance in such plants.

In trees, however, which may attain a height of 10 to 150 meters, the manner in which the necessary water-supply is carried to the leaves becomes a question of great complexity. Root-pressure is present in trees only during a limited period at the beginning of the growing season and is almost entirely absent in summer when the greatest amount of water is used. Hence it cannot bear a very important part in the ascent of sap. The transpiration of water from the leaves creates a vacuum in the stem below, as has been demonstrated in

Experiment 26. (See also Experiment 35.) The suction thus caused would not raise water higher than a suction-pump (about 10 meters). The water, however, is not in a continuous tube like the cylinder of a pump. The rectangular wood-cells are in the form of a series of chains. The water in each cell is separated from that of the neighboring cells by a thin membrane which promotes osmose. Water is transpired from the topmost cells of these chains, the cell-sap becomes concentrated and draws water from the cells beneath, and they in turn from those beneath them. There is thus formed a series of osmometers extending from the leaves to the roots, and capable of lifting water to any height.

In passing from the lower to the upper end of the narrow wood-cells, the ascent of sap is greatly retarded by capillary friction. On the other hand, the cavity of a wood-cell contains a bubble of gas, which by its expansion and contraction aids in forcing the sap upward. Further, *imbibition* by the cell-wall allows the passage from one part of the plant to another of a small amount of water which does not enter the cell-cavities.

It is difficult to account for the rapidity of sap-movements by the action of these physical forces alone. Some investigations tend to show that the protoplasm of the *wood-parenchyma* has a rhythmic osmotic attraction for water. Some such force is necessary to account for all features of sap-movement in trees.

EXPERIMENT 34.

AMOUNT OF WATER FORCED UPWARD BY ROOT-PRESSURE COMPARED WITH THAT TRANSPIRED BY THE LEAVES OF AN HERBACEOUS PLANT.

With a sharp knife cut off a strongly growing Sunflower plant near the ground. Fasten the upper part with its cut end in a measuring-cylinder containing water. To the stump (lower part) fasten by means of rubber tubing a tube bent twice at right angles. Insert the free end of the tube in a test-tube. The water thrown out and through this tube by root-pressure will be collected in the test-tube, and its volume can be compared with the amount drawn out of the measuring-cylinder by the transpiration of the other part of the plant (Fig. 25).

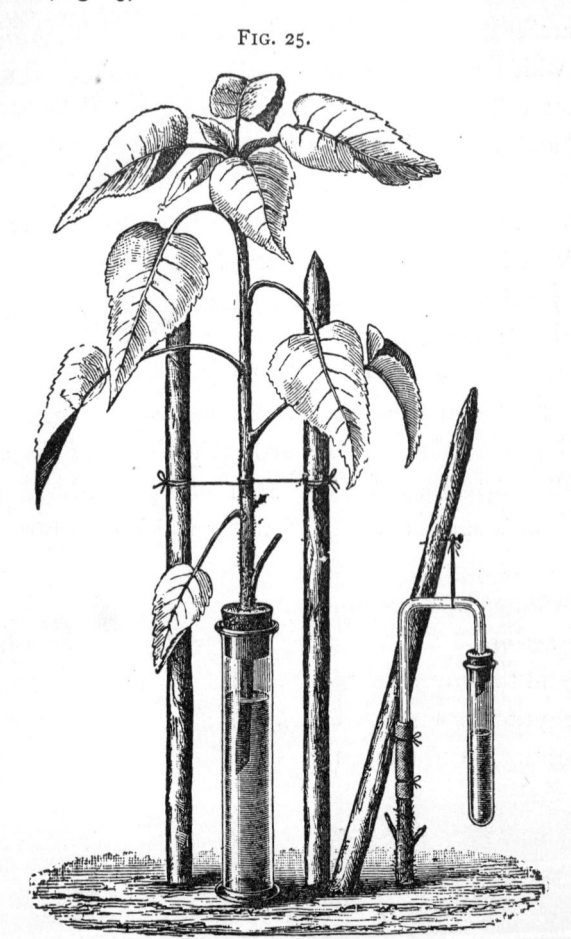

FIG. 25.

Comparison of root-pressure and transpiration. (After Oels.)

EXPERIMENT 35.

NEGATIVE PRESSURE.

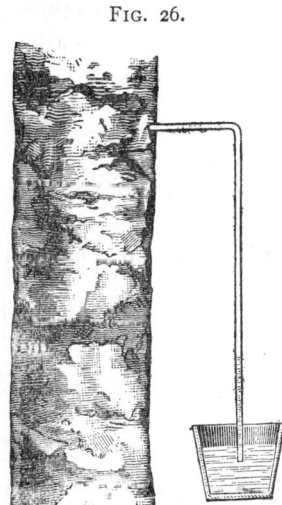

Fig. 26.

In September bore a small hole 6 cm. in depth in the trunk of a small Birch, and fit into the opening a glass tube a meter in length which has been bent once at right angles. Make the fitting "air-tight" by means of wax. Place the lower end of the tube in a dish of mercury. In a day or two the mercury will rise in the tube to a varying height. The rapid transpiration from the leaves has withdrawn so much water from the trunk of the tree that a partial vacuum is formed. (Fig. 26.) This may also be demonstrated as follows (Fig 27).

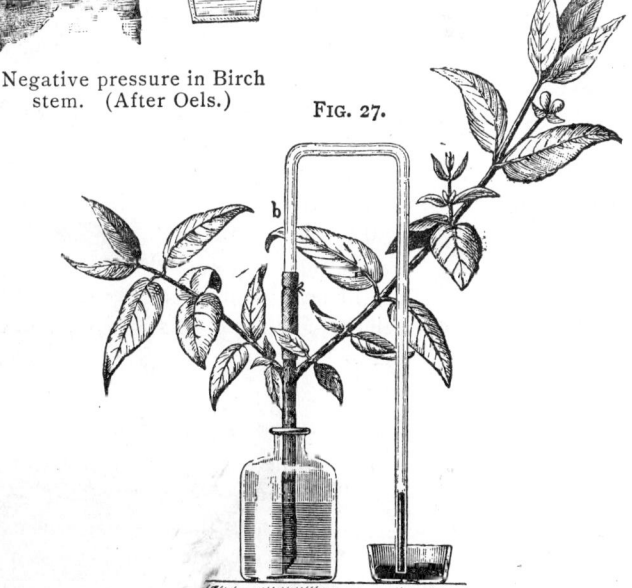

Negative pressure in Birch stem. (After Oels.)

Fig. 27.

Negative pressure in shoot of Lonicera. (Detmer.)

Cut off a shoot of some woody plant with tender leaves (Lonicera), and place the lower end in a vessel of water. Now cut off the top and fasten to the end of the shoot, by means of a piece of rubber tubing, a glass tube bent twice at right angles. Place the end of the perpendicular long arm in a vessel of mercury. In a short time the fluid ascends in the tube.

EXPERIMENT 36.

RESTORATION OF SAP-CURRENT.

Fix an excised shoot of Coleus or Helianthus (Sunflower) by means of a rubber stopper in one arm of a U tube and fill with water. Its power of conduction has been destroyed and it wilts (See Experiment 28). Now pour mercury into the free arm of the tube. The turgor is restored, and is retained until the mercury is higher under the plant than in the other arm.

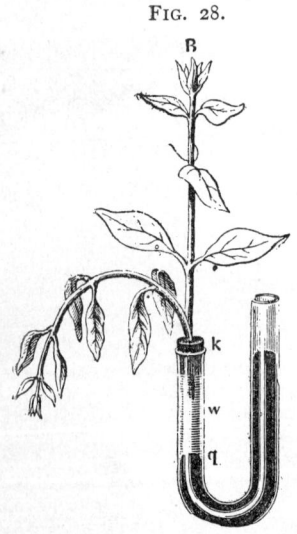

EXPERIMENT 37.

RATE OF ASCENT OF SAP.

Water copiously the soil in which a herbaceous plant 1 meter in height is growing, with a solution of lithium nitrate. In an hour cut a portion from the tip and at successive intervals toward the root. Burn these pieces in the flame of an alcohol-lamp or Bunsen burner, and by the characteristic red flame of lithium ascertain to what height the lithium has ascended in the stem.

FIG. 28.

Restoration of sap-current. (Sachs.)

EXPERIMENT 38.

MOVEMENT OF FLUIDS IN CONTINUOUS VESSELS.

Cut away the stem of a Euphorbia (Spurge), Sonchus (Wild Lettuce), or Asclepias (Milkweed). The milky juice exudes rapidly from both the upper and lower cut surfaces in a manner indicative of pressure. Cut away the stem of a Gourd or Pumpkin and note the large drops of slime which must have been forced from some distance, since that amount would not be found in the cells of the part cut across.

25. Path of Sap Movements.

The plant takes up water and mineral salts from the soil, and forms foods from carbon dioxide in the leaves. These substances must pass from the roots upward and from the leaf downward to be of use to the plant. The ascending stream moves upward through the woody part (*xylem*) of the stem. In trees the greater amount passes

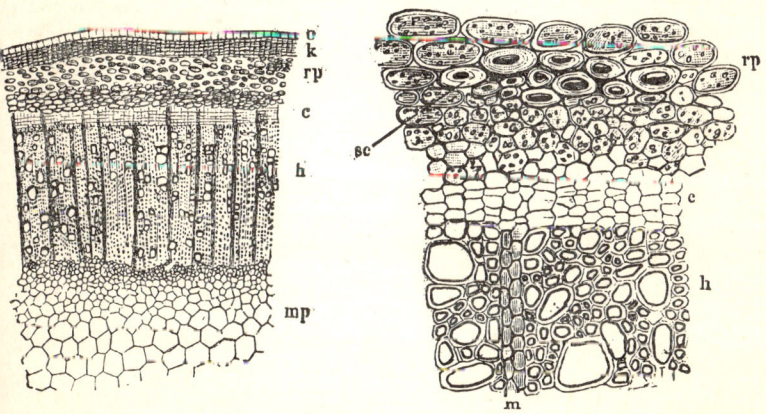

Fig. 29.

Fig. 30.

Cross-section of portion of shoot of Sambucus nigra (Elder) magnified 15 times. (After Oels.) *e*, epidermis; *k*, cork; *rp*, parenchyma and sclerenchyma; *c*, cambium; *h*, wood; *mp*, pith.

Cross-section of portion of stem of Sambucus nigra (Elder) magnified 150 times. (After Oels.) *rp*, phloem parenchyma; *sc*, sclerenchyma; *c*, cambium; *h*, wood; *m*, medullary rays.

through five or six of the recently-formed annual rings, as may be seen in trees with hollowed trunks which sustain tops of normal size. The descending current passes through the soft inner bark, the *phloem*. The descending current moves very slowly, and is carried on principally by diffusion (Figs. 29 and 30).

EXPERIMENT 39.
UPWARD PATH OF SAP.

Remove a ring of the bark and soft wood from any young tree or woody shoot a few centimeters above the ground by means of a sharp knife. The shoot shows no disturbance for a time varying from a few weeks to many months, when the roots become starved from lack of food usually supplied by the leaves and perish.

EXPERIMENT 40.

DOWNWARD PATH OF SAP.

In the same manner as above girdle a Willow branch 1 to 3 cm. in diameter by removing a ring of bark near the lower end.

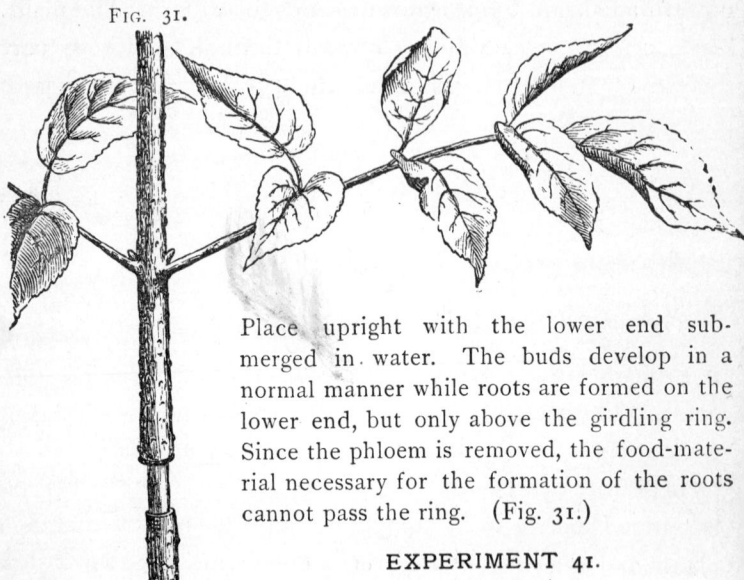

FIG. 31.

Place upright with the lower end submerged in water. The buds develop in a normal manner while roots are formed on the lower end, but only above the girdling ring. Since the phloem is removed, the food-material necessary for the formation of the roots cannot pass the ring. (Fig. 31.)

Girdled shoot of Sambucus. (After Oels.)

EXPERIMENT 41.

DEMONSTRATION OF PATH OF SAP BY COLORED FLUID.

Cut off a semi-transparent stem of Impatiens (Touch-me-not), and place the lower end in a water solution of some aniline color. In an hour note that the colored fluid has ascended in the woody fibres in the soft stem. Repeat, using a stalk of a young Corn plant. Allow it to stand in the solution 24 hours, then dissect and determine the path of the fluid.

CHAPTER III.

ABSOPTION OF GASES.

26. Gases used by the Plant.—Of the gaseous elements which enter into the food of plants, hydrogen is taken up in the form of water or ammonia by ordinary green plants, while it forms a large proportion of the complex substances which are used by parasitic and saprophytic plants. Oxygen is obtained from the air in a free state, and in combination in the form of water, carbon dioxide, and the mineral salts. Nitrogen is derived chiefly from compounds in the soil. Leguminous plants and many groups of the lower forms are able to take up this element directly from the atmosphere. The greater part of it used by the higher plants has been fixed in the soil by the action of Bacteria and related forms. At the present time the power of the various groups of plants to take up free nitrogen is not clearly defined. Carbon is obtained by plants which do not contain chlorophyll from the complex compounds which they use as food. This is true of all plants which use complex foods. Green plants, however, obtain their carbon supply from the carbon dioxide of the air. (§ 29.)

27. Diffusion of Gases.—If two gases that will mix are separated by a membrane, they will pass through the membrane by osmose in the same manner as liquids. The air is a mixture of 77.95 parts of nitrogen, 20.61 parts of oxygen, 1.40 parts of aqueous vapor, and .04 part of carbon dioxide. These gases are in different proportions in the plant and consequently a constant diffusion through the outer membrane takes place. Some cells containing substances which have a high osmotic equivalent for oxygen absorb it from the air. In like manner cells containing chlorophyll take up carbon dioxide during the daytime. Gases will readily diffuse through a membrane, yet cannot be forced through it by pressure.

EXPERIMENT 42.

DIFFUSION OF GAS THROUGH EPIDERMIS.

Smooth one end of a glass tube with an internal diameter of .5 cm. and a length of 30 cm. in a flame. Select a smooth and perfect grape. Take off the skin and clean the pulp from the inside. Place over the end of the tube, bringing the edges down and fastening closely to the tube by a small cord. (Fig. 32.) With sealing-wax secure the edges to the glass in such a manner as to be "airtight." Test by placing in water and forcing air in at the other end. If no bubbles escape, fill the tube with water and invert in a vessel of mercury. Displace the water with carbon dioxide and note the height of the mercury column daily for a month. By the diffusion of the carbon dioxide through the membrane the column of mercury may be raised as high as 26 cm.

Remark.—In inverting the tube when full of water no air must be allowed to gain entrance. To obtain carbon dioxide use the apparatus described in Experiment 57. Marble and hydrochloric acid should be used instead of zinc and sulphuric acid, as there described.

28. Absorption of Hydrogen, Oxygen, and Nitrogen. — The absorption of hydrogen, oxygen, and nitrogen, and their synthesis into food are so closely connected with other complex metabolic processes that a consideration of the separate action in each case is somewhat difficult. The manner in which carbon is obtained and used is, however, a fairly distinct process.

29. Photosynthesis.

Green plants absorb carbon dioxide from the air either through the epidermis or the stomata. Carbon dioxide is composed of one part of carbon and two parts of oxygen. The protoplasm which forms the mass of the green color bodies (chlorophyll bodies) in the cells has the power, when it receives the sunlight, of separating one part of the oxygen which is thrown off as a free gas, while the carbon monoxide which remains is combined with the water present to form a compound of carbon, hydrogen, and oxygen from which sugar is ultimately derived. The entire process may be designated *photosynthesis*. No life is imaginable without photosynthesis. All organisms, plants, and animals alike are ultimately dependent upon the products of this process for their carbon compounds.

FIG. 32.

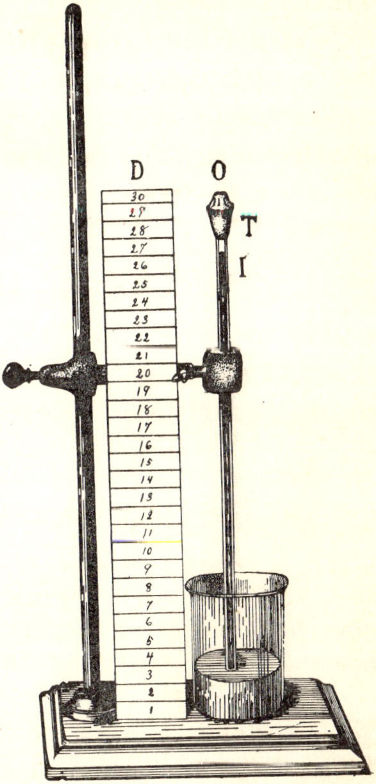

Diffusion of gas through epidermis. *I*, level of mercury column 20 days after beginning of experiment; *O*, skin of grape; *T*, sealing-wax; *D*, centimeter-scale.

EXPERIMENT 43.

THE ACTION OF LIGHT IS NECESSARY FOR PHOTOSYNTHESIS.

Weigh 4 seeds of Corn, germinate and grow in nutrient solution. Place 2 of the seedlings in a dark chamber and the remaining 2 in the sunlight. In three weeks take the plants from the solutions, dry in the air for several days and weigh. Those in darkness will have lost, while those in light will have gained weight since they were able to form food from carbon dioxide of the air and water.

EXPERIMENT 44.

OXYGEN IS GIVEN OFF DURING PHOTOSYNTHESIS.

Fill a funnel of medium size with green shoots of Elodea 8 to 10 centimeters in length. Immerse the funnel, inverted, in a wide dish filled with spring-water, and over the small end of the funnel place a test-tube filled with water. Set in the sunlight. In a short time

FIG. 33.

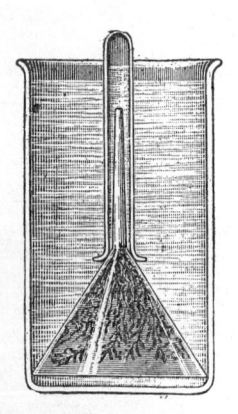

FIG. 34.

Apparatus to show excretion of oxygen by Elodea. (Detmer.)

Action of light on Elodea. (After Oels.)

gas can be seen, collected in the upper part of the tube, which tested with a glowing splinter is proved to be oxygen.

EXPERIMENT 45.

THE AMOUNT OF PHOTOSYNTHESIS AND OF OXYGEN GIVEN OFF DEPENDS ON THE INTENSITY OF THE LIGHT.

Fasten a shoot of Elodea about 10 centimeters long to a glass rod and immerse in spring-water or water containing carbon dioxide, so that the cut end is higher than the other. Set the apparatus in direct sunlight, and immediately a stream of gas-bubbles—oxygen— begins to pour from the cut end of the shoot. (Fig. 34.) In diffused light, or in light the intensity of which is reduced by means of one or more plates of ground glass (Fig. 35), the number of bubbles

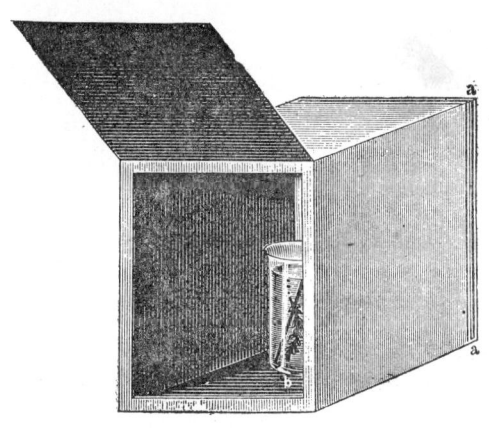

FIG. 35.

Box blackened on the inside. (After Oels.) *a a*, ground-glass plates; *b*, shoot of Elodea.

given off decreases, so that the dependence of photosynthesis upon light can be seen directly.

30. Physical Properties of Chlorophyll.—Chlorophyll is a substance of extremely complex and unstable constitution. It is generally found in certain definite masses of protoplasm in the cell, although in some plants it appears uniformly diffused throughout. Its presence is sometimes masked by other coloring matter, as in the Red Sea-weeds and colored leaves of the foliage plants of the garden. Some of the autumnal tints of leaves are due to coloring substances resulting from the oxidation of chlorophyll. The spectrum of sunlight which has passed through a solution of chlorophyll in alcohol shows several dark bands. The portions of light thus absorbed are converted into heat and other forms of energy needed for photosynthesis and other processes. If different portions of the spectrum are allowed to act on a plant, the relative amount of photosynthesis promoted by each can be demonstrated. The red rays are principally active in photosynthesis.

31. Division of the Spectrum.—It is found that a watery solution of potassium bichromate transmits only the red, orange, and yellow rays of light, and that an ammoniacal solution of copper oxide transmits only the blue and violet rays. Unless carefully compounded, however, the latter solution will also allow the passage of some of the red and yellow rays.

EXPERIMENT 46.

PORTION OF THE SPECTRUM ACTIVE IN PHOTOSYNTHESIS.

Make a solution of potassium bichromate in water. To obtain the ammonia-copper-oxide solution, add ammonium hydrate to a solution of copper sulphate in water as long as the forming precipitate is redissolved. Fill a double-walled bell-jar with each solution.

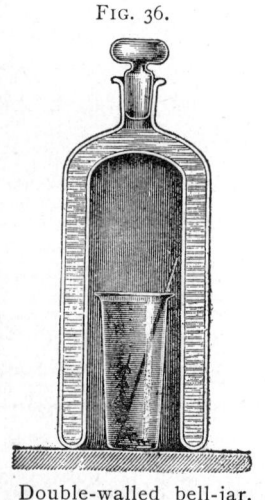

FIG. 36.

Double-walled bell-jar. (After Oels.)

FIG. 37.

Apparatus to replace double walled bell-jar. (After Oels.) k, solution of potassium bichromate or copper oxide; w, water; p, pasteboard cover.

Prepare three shoots of Elodea as in Experiment 45. Place one in sunlight and one under each bell-jar. Bubbles of oxygen are given off from the shoots in open sunlight and under the red bell-jar, but none from the one under the blue bell-jar. (Fig. 36.)

If the double-walled bell-jars are not at hand, each may be replaced by two glass cylinders, one so much larger than the other that when the smaller is fastened inside the other by means of a cork, a space of about 1 to 2 cm. remains between them. Fill this space with the proper solution, and the inner vessel with water containing carbon dioxide, and place in the latter the plant-shoots. To cut off the perpendicular rays cover the apparatus with a loosely-fitting cardboard cover. (Fig. 37.) The pasteboard box shown in Fig. 35 can also be used if instead of the ground-glass plates (*a a*) parallel-walled glass cells filled with the absorption fluid are used. With colored glass plates only an approximately pure light can be obtained. If the box is used, it will be found most convenient to place the shoot in an inverted test-tube filled with water as in Fig. 38. The amount of gas can be measured directly in the tube.

FIG. 38.

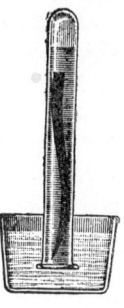

Apparatus to demonstrate the excretion of oxygen. (Mangin.)

32. Product of Photosynthesis.—The product of photosynthesis is probably some soluble carbohydrate such as glucose. As soon as enough of this substance has been formed to meet the immediate needs of the plant the remainder is converted into starch. If the plant is placed in darkness or under any condition in which it cannot carry on photosynthesis, as soon as the glucose in the cells is consumed the starch is reconverted into glucose or some form of sugar and assimilated. The amount of starch present in a plant may be taken as an indirect indication of the amount of photosynthesis.

EXPERIMENT 47.

MACROCHEMICAL TEST FOR STARCH.

Boil a few leaves of Bean, Tomato, or Tropæolum for a few minutes to kill the protoplasm and swell the starch-grains present. Place in warm alcohol until the chlorophyll is dissolved. Bring the leaves into an alcoholic solution of iodine for a half-hour. The leaves will be colored a dark blue if they contain starch.

EXPERIMENT 48.

MICROCHEMICAL TEST FOR STARCH.

Decolorize some filaments of Spirogyra, or leaves of Funaria as above, and place on a glass slide in a drop of chloral hydrate (chloral hydrate 5 parts, water 2 parts). Add a drop of a solution of iodine in iodide of potassium, and examine with the microscope. A thin section of large leaves may be examined in this manner.

EXPERIMENT 49.

STARCH AS AN INDICATION OF PHOTOSYNTHESIS.

Place some Spirogyra or Vaucheria in a dark chamber for 24 hours. Test some of the filaments for starch. It will be found absent. Set the vessel containing the filaments in the sunlight for a few minutes. Examine a second lot. Starch will be found present.

EXPERIMENT 50.

FORMATION OF STARCH FROM SUGAR.

Deprive a Geranium plant of starch by placing in a dark chamber for 24 hours or longer. Test for starch, and if none is present cut off a leaf and place it in a 20% solution of sugar for a week in a dark chamber. Test for starch. The protoplasm of the leaf has used the sugar as food and has also converted a portion of it into starch.

CHAPTER IV.

RESPIRATION AND OTHER FORMS OF METABOLISM.

33. Nature of Metabolism.—By various processes, of which photosynthesis is an important example, a large number of complex substances are formed in the plant. The synthesis of complex compounds from those of simpler composition is termed *constructive metabolism*. In this process a portion of the oxygen in the simple compounds is liberated. Thus in photosynthesis water and carbon dioxide, both containing oxygen, are combined and a portion of the oxygen is set free. This liberation of oxygen makes constructive metabolism what is known in chemistry as a *reducing process*. On the other hand, when the complex foods thus formed are used by the plant, oxygen is taken up and the complex substances are resolved into others of simpler composition. This is known as *destructive metabolism*. Since oxygen is absorbed in destructive metabolism, it is essentially an *oxidizing process*. One of its most important forms is respiration.

34. Respiration.—In respiration, which is directly opposite in character to photosynthesis, oxygen is absorbed, carbon dioxide given off, and energy liberated in the forms of heat and electricity. The oxygen needed is largely obtained from the air, although in some instances it is derived from other compounds in the plant which contain a large porportion of it. The extraction of oxygen from one substance within the plant for the oxidation of another is termed *intramolecular respiration*. This form of respiration is carried on to some extent by all plants, but is characteristic of the germination of oily seeds and of Yeast, Bacteria, etc.

Respiration is essentially the same process in both plants and animals, but while the former breathe and give off carbon dioxide constantly from all parts of their bodies, the latter, in the highly-developed forms, breathe rhythmically and for the greater part by means of organs especially adapted for the purpose. Yet there are instances among both plants and animals where the respiratory processes are suspended or reduced for a period of varying length.

35. Absorption of Oxygen and Excretion of Carbon Dioxide.—The amount of carbon dioxide given off in respiration is

FIG. 39.

Liberation of carbon dioxide by respiration. (Mangin.) *a*, baryta-water.

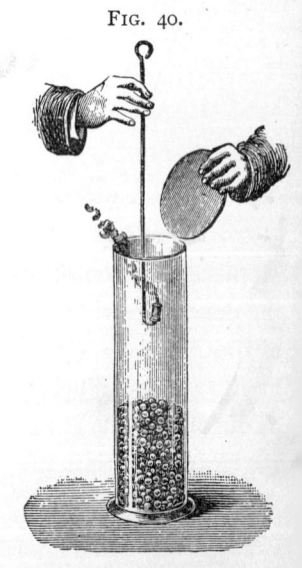

FIG. 40.

Cylinder containing germinating Peas. (Sachs.)

approximately equal to the oxygen taken up. The proportion of the two substances varies with the temperature and other conditions, and in the different organs. De Saussure found that 1 gram of seed of Hemp absorbed 19.7 cc. of oxygen and exhaled

13.26 cc. of carbon dioxide in the same time, and 1 gram of seed of Madia absorbed 15.83 cc. of oxygen, while it exhaled 11.94 cc. of carbon dioxide. Young growing plants will exhale an amount of oxygen equal to their own volume in 24 to 36 hours.

EXPERIMENT 51.

EXCRETION OF CARBON DIOXIDE BY LEAVES.

Provide a ground-glass bell-jar and plate. Under the bell-jar place a well-leaved plant grown in a pot, and a vessel containing lime- or baryta-water; place the apparatus in darkness. After a short time a film of carbonate can be seen on the surface of the fluid which, if allowed to remain longer, collects as chalk (or baryta). As a control experiment, set up the same apparatus without the plant. The lime- or baryta-water is scarcely affected. (Fig. 39.)

EXPERIMENT 52.

EXCRETION OF CARBON DIOXIDE BY GERMINATING SEEDS.

Fill a glass jar of 1 liter capacity one-third full of Peas which have lain a day in water. Cover tightly. After 12 or 14 hours a light thrust in is extinguished, showing the lack of oxygen, and a vessel containing lime- or baryta-water placed inside demonstrates the presence of carbon dioxide. Instead of Peas, developing heads of a Composite or some large Fungus can be used. (Fig. 40.)

36. Liberation of Heat.—In very strong respiration, as in the development of flower-heads of the Compositæ, flower-tubes of the Aroids, and germinating seeds, enough heat is liberated in the combustion of the carbon compounds of the plant to be easily detected by the thermometer. Sachs observed in 100 to 200 germinating Peas a rise in temperature of 1.5° C. (Fig. 41.)

EXPERIMENT 53.

HEAT LIBERATED BY GERMINATING SEEDS.

Fill a glass funnel of medium size with germinating Peas or blooming heads of Leontodon, Anthemis, Bellis, etc., into which a thermometer graduated to $\frac{1}{5}$ degree C. has been thrust. To avoid loss of heat as far as possible, cover the funnel with a perforated glass plate, whereby access of air is prevented. The carbon dioxide formed is absorbed by a solution of potassium hydrate which is placed in a glass dish under the funnel. As a comparison place near this apparatus a thermometer in the free air which will not be affected by the heat of the plants. To obtain the most uniform temperature for both thermometers cover each with a large bell-jar.

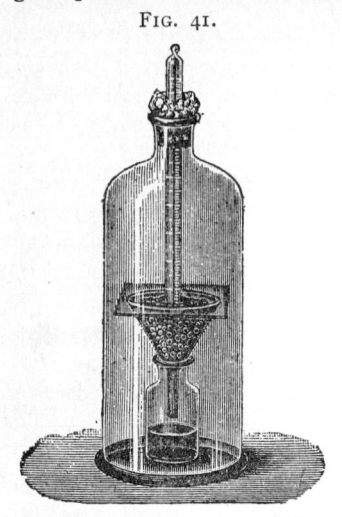

FIG. 41.

Apparatus to demonstrate liberation of heat in respiration. (Sachs.)

37. Respiration, Essential to Growth and Dependent on Air.—The conversion of food into living substance is possible by means of respiration only. The higher plants may carry on a certain amount of intramolecular respiration and thus accomplish a small amount of growth. For the normal development of the plant, however, it is necessary that it have access to the free oxygen of the air.

EXPERIMENT 54.

OXYGEN NECESSARY FOR RESPIRATION.

Fill two respiration-tubes of 100 CC. capacity with water which has been boiled to drive off the dissolved air. In the bulbs of each insert a half-dozen seeds of Pea or Wheat, and invert over a dish of mercury. Twenty-four hours later displace nearly all of the water in one tube with hydrogen and the other with air. The seeds in hydrogen do not germinate, while those in air, which are able to obtain their customary supply of oxygen, develop normally. To obtain the hydrogen, place a few grams of granulated zinc in a flask or bottle, and cover to a depth of 5 cm.

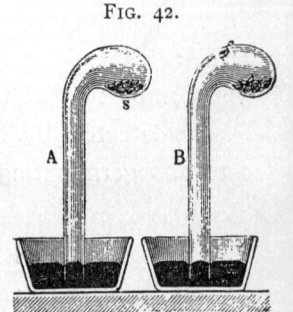

FIG. 42.

Respiration-tubes. (Detmer.) A, filled with hydrogen, and B, oxygen.

with diluted sulphuric acid. Close the mouth of the flask with a cork stopper through which extends a short section of glass tubing. To the outer end of this attach a section of rubber tubing 30 cm. in length. The free end of the rubber tube should be fitted with a small piece of glass tubing drawn to a point and bent at an angle of 45 degrees for introducing the gas into the respiration-tube. (Fig. 42.)

38. Fermentation.—Perennial plants which grow in temperate climates store up a supply of reserve food in the roots, rhizomes, or stems, to serve as building material at the beginning of the next vegetative period. The seedling cannot obtain nourishment from the soil and air during the first period of its development, because its roots are not sufficiently developed, and because it has not yet enough chlorophyll to build up food, by aid of the sunlight. Before the solid reserve substances can become of use to the plant they must be dissolved, and transported by diffusion where they are needed. The solution of the reserve food is accomplished by means of ferments or *enzymes*, which by their presence induce changes in organic compounds (*fermentation*) without themselves being thereby in any way affected. On account of this last property a small amount of enzyme may cause fermentation in a large quantity of the substance acted upon. In the germination of a seed, external moisture and temperature stimulate the protoplasm to form an enzyme which dissolves the solid starch, protein, or fat in the storage cells. The solution is diffused into the growing cells of the young plant where it is used in building up protoplasm. The starch formed in leaves undergoes solution and transportation in a similar manner. *Diastase*, the enzyme which changes starch into maltose, is perhaps the most widely distributed ferment.

EXPERIMENT 55.
ACTION OF DIASTASE.

Place 10 grams of seed of Barley in a germinator for 36 hours, or until the radicles are .5 cm. in length. Grind fine, in an ordinary coffee-mill, and add to three parts of water. After a time filter and mix the filtrate, which now contains diastase, with a fifth part of very thin starch paste (1 gram starch, 100 grams water). A sample of this mixture is colored blue on treatment with iodine, a sample taken later, violet, then brown, and finally one taken after two or three hours is colorless, demonstrating that all the starch has been transformed into maltose or sugar by the diastase present in the germinated seed.

Cut thin sections of the seeds at the beginning of the experiment and determine the appearance and characteristics of the starch-grains. Make a similar examination 24 and 48 hours later. Allow germination to proceed in a few seeds, and examine 4 days later. The starch-grains are gradually corroded and dissolved by the diastase formed.

EXPERIMENT 56.
TRANSLOCATION OF STARCH.

A Tropæolum plant whose leaves are rich in starch is placed in the dark after some of its leaves have been cut off. The excised leaves are likewise placed in the dark in a moist room or under a bell-jar. After a few days test some of the excised leaves and those remaining on the plant for starch. Those on the plant show some starch, mostly in the nerves, while those excised show starch in the other parts as well, because they could not transfer it to other organs.

EXPERIMENT 57.
FORMATION AND TRANSLOCATION OF STARCH.

On a well-developed plant of Tropæolum majus standing in the sunlight in the forenoon, darken portions on some healthy leaves, by means of cork plates, fastened on opposite sides by pins. On the afternoon of the following day cut off the leaves and boil in water in a porcelain dish for a few minutes, to kill the protoplasm. Extract the coloring matter by alcohol many times renewed. The decolorized leaves are now saturated with alcoholic iodine in a porcelain dish, whereupon they will be colored a deep blue except in shaded portions. Since, substantially, starch-formation in the leaf proceeds by day only, and the solution and translocation at night as well, the places exposed to the sunlight contain enough starch to

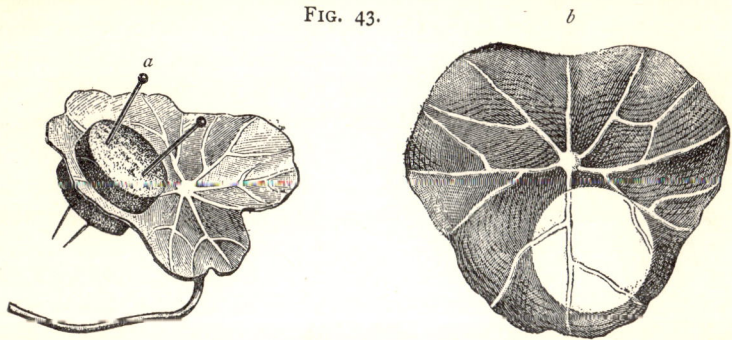

FIG. 43.

a, Tropæolum leaf to which are attached two pieces of cork to prevent photosynthesis. (Detmer.) *b*, same after removal of cork, treated with iodine.

give the microscopic reaction with iodine. It was taken away from the shaded places at night, however, and could not be replaced. (Fig. 43.)

39. Changes in Color.—Many changes in the chemical composition of substances in the plant are accompanied by corresponding changes in color. (See Chlorophyll, Par. 30.) Flowers which are blue when fully opened were originally red in the bud. The sap was acid at first and became alkaline as the result of metabolic changes. Leaf-colors offer similar conditions, although the changes in color here are sometimes due to the oxidation of chlorophyll and other coloring matter in the cells.

EXPERIMENT 58.
RELATION OF RED AND BLUE COLORS OF FLOWERS.

Immerse a leaf of Begonia bearing red hairs, for a short time in a weak solution of ammonia. The hairs become blue. Place a blue petal of Myosotis in a 1% solution of acetic acid. It becomes reddish. Express the sap from a handful of petals of Roses or Peonies. Collect in a test-tube and add a few drops of ammonia. A blue color results. Add some acid. The red color is restored. Observe the different colors assumed by leaves in the autumn.

CHAPTER V.

IRRITABILITY.

40. Nature of Irritability.—The term *irritability* designates that property of plants by which they respond to certain influences known as *stimuli*. The stimuli may be either internal forces set in operation by metabolic activity or external influences, such as gravity, light, temperature, electricity, moisture, and mechanical contact. The plant may react in two ways, first, by changes in the structure, form, and size of its organs; second, by motion or change in position of its organs or of the protoplasmic bodies in its cells. The reactions of the first class concern growth; those of the second class result in placing plant or cell organs in certain positions relative to the direction of the stimulus. Thus a plant grown in darkness develops its stems and leaves in quite different form and structure from one grown in the open air (Fig. 67). Light, then, affects the structure and form of plants by what is known as its *formative* or *tonic* influence. (See Chapter VI.) Light also causes shoots to bend toward its source in such manner as to place their axes parallel to the light-rays. This is termed its *directive* influence. A stimulus may give rise to reactions of both kinds at the same time, and they cannot always be easily and distinctly separated by experiments.

41. Perceptive and Motor Zones.—The action of an external stimulus on one part of a plant does not necessarily cause a movement in that part, but the impulse may be transmitted to a region more or less distant. Thus, for instance, a touch on the leaf-blade of a Mimosa (Sensitive-plant) causes no contraction in the leaf-blade, but the impression is transmitted to the pulvinus at the base of the leaf or leaflet and produces movement. The region which receives the stimulus is designated the *perceptive zone*, and the one which causes the movement, the *motor zone*. The two may coincide in position.

42. Geotropism.—The power by which a plant responds to the influence of gravity is termed *geotropism*. The response of an organ to this stimulus may occur in three ways, as follows. (1) The organ may point its apex toward the centre of the earth, the source of gravity, in which instance it is said to be *progeotropic*. This action is generally manifested by primary roots. (2) It may point its axis away from the source of gravity, directly upward, when it is said to be *apogeotropic*. Erect shoots are generally apogeotropic. In general organs of radial structure exhibit one of these two forms of geotropism. (3) The organ may place its axis in a horizontal position, at right angles to the force of gravity, when it is said to be *diageotropic*. This is characteristic of the larger number of bilateral organs, such as leaves, although also shown by organs of radial structure, such as branches of stems, secondary roots, etc.

EXPERIMENT 59.

PROGEOTROPISM.

To a cork in the top of a bell-glass fasten a seedling of a Bean with the radicle which is 1 to 2 cm. in length in a horizontal position. In a few hours the tip is found to be pointing downward more or less directly. (Fig. 44.)

EXPERIMENT 60.

GEOTROPIC REACTIONS OF SEEDLINGS.

Place a layer of sawdust between two horizontal parallel rings covered with wide-meshed gauze.

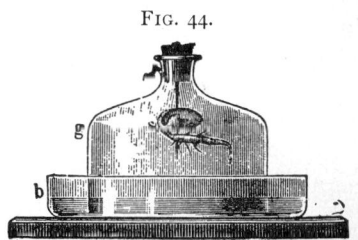

FIG. 44.

Progeotropism of seedling. (Detmer.) *b*, dish partly filled with water; *g*, bell-glass; *s*, seedling.

Plant seeds of the Pea, Bean, or Corn in the sawdust. The roots pass through the meshes downward and the shoots upward. Invert the apparatus and these organs will bend in the opposite directions. (Fig. 45.)

FIG. 45.

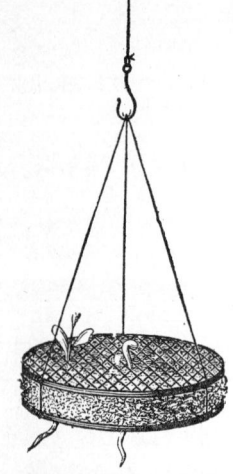

FIG. 46.

Geotropism of roots and shoots. (After Oels.)

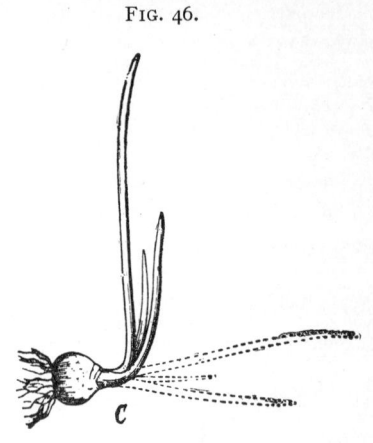

Apogeotropism of leaves of Onion. (Frank.)

EXPERIMENT 61.

APOGEOTROPISM.

Place a Tulip, Hyacinth, Onion, or Fritillaria, which is growing rapidly, in a horizontal position. In a short time the leaves curve directly upward. (Fig. 46.)

EXPERIMENT 62.

DIAGEOTROPISM.

Observe the opening flower-buds of a Narcissus, which at first are erect, but later the perianth-tube assumes a horizontal position. After they have attained this position lay the pot on its side with the leaves and stems horizontal and the perianth-tube pointing downward. In 10 hours the pedicels will have again curved to place the perianth-tube in the same position as before.

43. Perceptive and Motor Zones of Roots.—The stimulus of gravity is received by a sensitive portion near the tip of a root (the perceptive zone), and an impulse is conveyed to a region several millimeters distant which curves (the motor zone).

IRRITABILITY. 53

The motor zone in this instance is located in the region of most active growth. (See Experiment 86.)

EXPERIMENT 63.
PERCEPTIVE ZONE OF ROOTS.

Repeat Experiment 59 after cutting away a portion of the tip of the root 1 to 2 mm. in length. The root does not now respond to gravity, and shows no movement until the tip is rehabilitated, when it curves downward in a natural manner.

EXPERIMENT 64.
MOTOR ZONE OF ROOTS.

With a fine brush carefully mark off equal spaces (2 to 3 mm.) on the primary root of a seedling of Phaseolus (Bean). Suspend in a horizontal position in a moist chamber. In a day note the region in which curvature has occurred, and its distance from the tip.

FIG. 47.

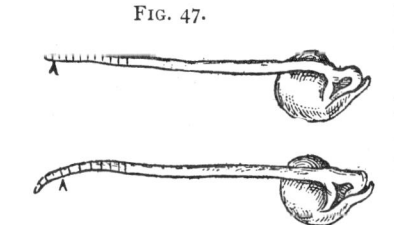

Motor zone in roots. (Pfeffer.)

EXPERIMENT 65.
MOTOR ZONES OF CULM OF GRASS.

Cut a length of 12 cm. from a vigorously growing culm of Grass. Place in a horizontal position in a moist chamber with one end imbedded in sand. Six hours later note the region of curvature. The motor zone will be found in the pulvinus like internodes. (Fig. 48.)

FIG. 48.

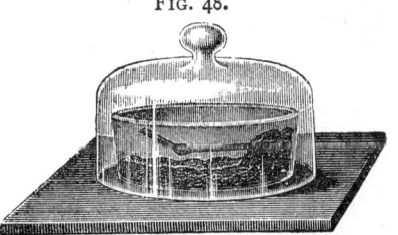

Curvature of Culm of Grass. (After Oels.)

EXPERIMENT 66.
FORCE OF CURVATURE.

Fasten a seedling of Pea with a radicle 1 cm. in length to a piece of cork attached to the side of a vessel containing mercury. Place the seedling in such position that the root is horizontal and the tip is in contact with the mercury. Pour in enough water to form a thin layer on top of the mercury. The root will bend downward with such force as to penetrate the mercury. (Fig. 49.)

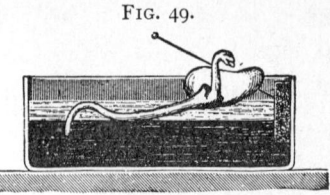

FIG. 49.

Root of seedling penetrating mercury. (Sachs.)

44. Influence of Gravity. — Gravity acts in a vertical direction and with a force directly proportioned to the mass of the body acted upon. In plants the amount and rapidity of the curvature of an organ, in response to gravity, depends on its stage of development and the angle which its axis forms with the vertical. Progeotropic organs respond most rapidly

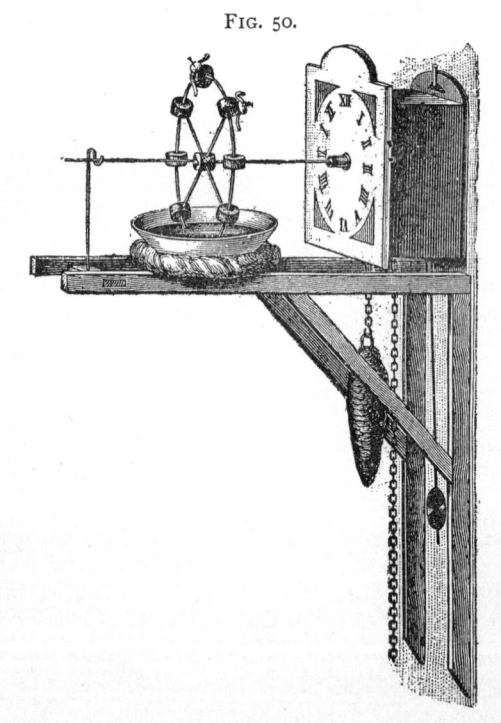

FIG. 50.

Seedlings on a revolving wheel driven by a clock. (After Oels.)

when their tips are pointing upward at an angle of 45 degrees from the vertical; apogeotropic organs respond most readily when pointing downward at an angle of 45 degrees; and diageotropic organs respond with equal facility in either position. The action of gravity upon a plant may be neutralized by placing it on the periphery of a wheel revolving in a vertical plane, or by turning the plant on its own axis in a horizontal position.

EXPERIMENT 67.

NEUTRALIZATION OF INFLUENCE OF GRAVITY.

Take the hands from a large wall-clock whose dial is parallel to the rays of sunlight (to avoid the disturbing action of light) and fasten to the prolonged axis a cork plate 10 cm. in diameter, or a wheel to the periphery of which are attached pieces of cork. Fasten seedlings of Pea to the cork in various positions by means of pins, and set the clock in motion. Twenty-four hours later each seedling will be found to be growing in the position in which it was placed, and no marked curvatures in any of the organs can be noticed. (Fig. 50.)

Remark.—The revolving wheel must be partially immersed in water or placed in a spray to keep the seedlings moist. A small American clock can be used instead of the large clock shown in the figure.

45. Replacement of Gravity.—When gravity is overcome by another force, the seedling tends to place its axis parallel to this new force in the same manner as toward gravity in its normal position. The force which acts upon a plant fastened to a rapidly-revolving wheel acts in a tangential direction; consequently the plant tends to place its axis parallel to the tangent. This is true of plants rotated in a vertical plane at a speed of 100 to 300 revolutions per minute with a wheel having a radius of 6 to 20 cm. If a plant is rotated in a horizontal plane at this speed, centrifugal force tends to cause the shoot axis to lie in a horizontal plane, while gravity tends to cause the axis to take a vertical position. In this instance the axis will take a position between the direction of these two forces,

The roots will point outward and downward, and the shoots upward and inward. These positions will be taken by rapidly-growing seedlings in 5 or 6 hours.

EXPERIMENT 68.

REPLACEMENT OF GRAVITY BY CENTRIFUGAL FORCE.

If a water system is at hand, the rapid rotation of the wheel or disk holding seedlings may be accomplished by the following apparatus

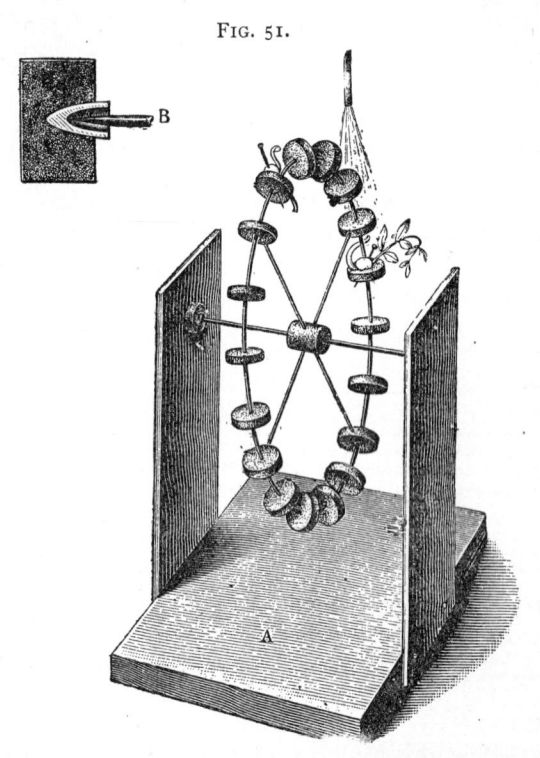

FIG. 51.

Centrifugal apparatus. (After Oels.) *A*, heavy board base ; *B*, cork holding the sealed end of a glass tube which serves as a bearing for the end of the axis of the wheel.

and the seedlings may be kept moist during the experiment. Upon a wire, 40 centimeters long, the size of a knitting-needle, are strung at equal distances a number of circular cork plates, 3 cm. in diameter and a half-centimeter thick. The wire is now bent in the form of a circle, the ends brought together through a piece of cork and united. Four brass wires serve as spokes, while the hub is made from a heavy cork. The axis is also made from a piece of wire, about 15 cm. in length, in order that the seedlings may have enough space for their curvatures. The axis rests in bearings made of the ends of sealed glass tubes (Fig. 51, B), which are fastened in stationary corks by means of sealing wax. If now a sufficient stream is allowed to fall perpendicularly on the cork plates on one side, a rapidity of revolution will be secured that will in five or six hours effect a noticeable change in position of the seedlings, which have been placed with their roots toward the center. If large casks are at hand, the experiment may be carried on without a water system. A glass tube of an internal diameter of 4 mm. with a fall of 1 meter of the water will furnish a stream that will revolve the wheel more than twice per second, which is entirely sufficient for the experiment. The replacement of the water is necessary for the continuation of the experiment.

46. Heliotropism, Thermotropism, etc.—Radiant energy in the form of heat, light, and electricity exercises a very marked directive influence on the position of plant-organs. The effect of sunlight is much better known than that of the other stimuli acting on the plant. It is a matter of general observation that the shoots of a large number of plants bend toward the light. A close inspection shows that various organs respond in a different manner to light, as also to gravity. Thus many roots direct themselves away from the source of light (*apheliotropism*), trailing shoots, leaves, etc., at right angles to the rays (*diaheliotropism*), while, as noted above, others, such as stems, bend toward the light (*proheliotropism*). It will be seen that while the force of gravity acts always in the same direction, the line of light-rays from the sun moves through an angle of 180 degrees daily. This change of the position of the source of light causes corresponding movements in heliotropic organs. Sunlight affords two separate stimuli to the plant: one from the blue-violet rays which causes heliotropic movements, and one from the red end of the spectrum which causes heat or thermotropic movements, as may be shown by Experiment 72. The purpose of the heliotropic as well as all other movements of plants is doubtless that of plac-

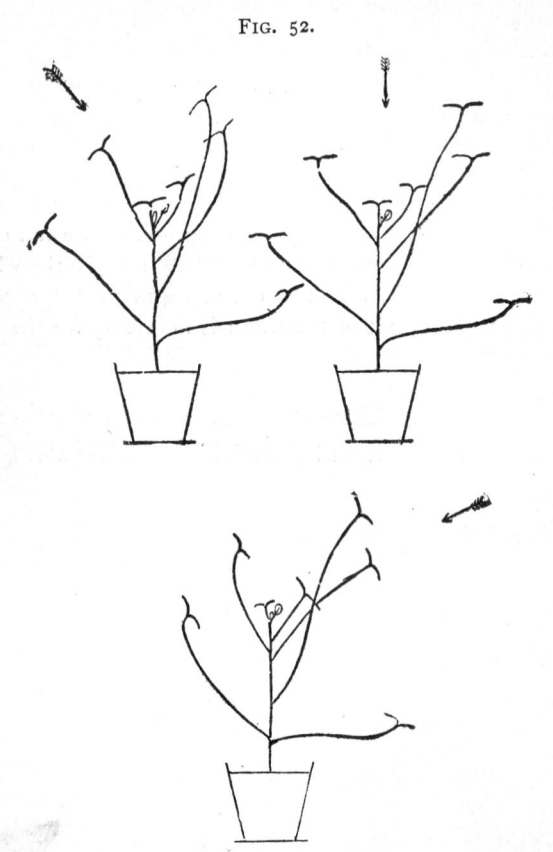

FIG. 52.

Diagram of light positions of leaves. The arrows denote the direction of the rays. (Vöchting.)

ing the plant-organ in the position best suited to the performance of its functions. Heliotropic movements place the leaves in a position most favorable for photosynthesis and transpiration.

IRRITABILITY.

EXPERIMENT 69.

PROHELIOTROPISM.

Place a Malva or Helianthus grown in a pot in the open air, near a window with a southern exposure. The leaves gradually assume the definite positions shown in Fig. 52.

EXPERIMENT 70.

HELIOTROPIC MOVEMENTS OF ROOTS AND SHOOTS.

Fasten seedlings of Sinapis alba (Mustard) or Phaseolus multiflorus (Bean) on a piece of tulle stretched lightly across a glass vessel filled with spring-water. After the roots and stems have attained a length of 1 cm. place the apparatus under a pasteboard box lined with black paper, through which the light may gain entrance by a small aperture. In a few hours the roots and stems will be influenced as described above. (Figs. 53 and 54.)

FIG 53.

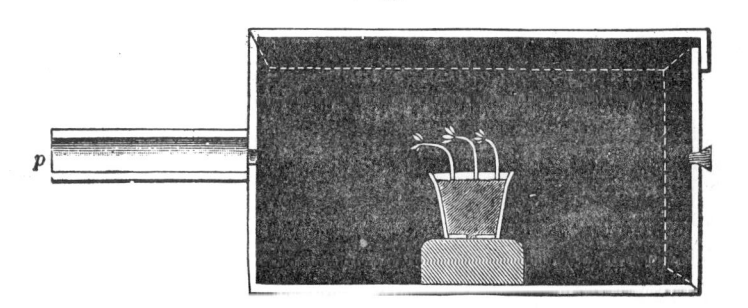

Dark chamber with a tube opening in one end. (Schleichert.)

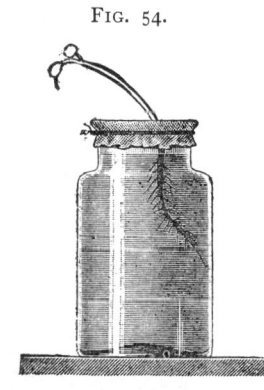

FIG. 54.

Seedling of Mustard grown under one-sided illumination. (Detmer.)

EXPERIMENT 71.

HELIOTROPIC MOVEMENTS OF LEAVES.

Bend and fasten in a horizontal position an upright well-leaved branch of the Maple, or a whole Helianthus plant, in the open air. Soon the leaves which were previously horizontal, and perpendicular to the shoot on all sides, show peculiar torsions of the petioles, which, so far as they are capable of growth, finally result in the placing of the leaves in a horizontal position, but parallel to the shoot and one another. (Fig. 55.)

EXPERIMENT 72.

EFFECT OF RED AND BLUE LIGHT.

Of two equally sensitive seedlings, place one in a chamber (Fig. 55), with one side of yellow glass, and the other in a similar chamber, with one side of blue glass. The heliotropic movement is much more marked in the blue light. Instead of the colored plates, use glass vessels with parallel walls, filled, one with a solution of bichromate

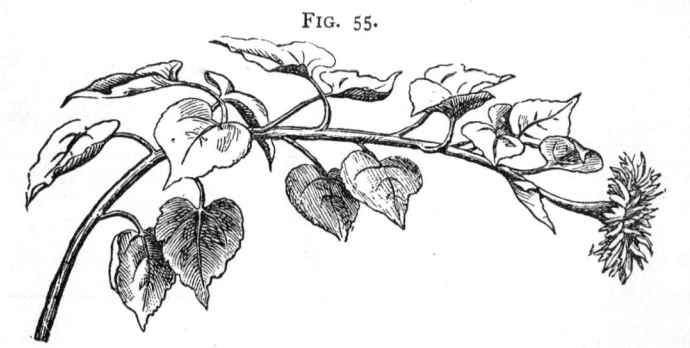

FIG. 55.

Shoot of Sunflower which has been in a horizontal position several days. (After Oels.)

of potassium, the other with a solution of ammonia-copper-oxide. Only small plants can be used in the experiment.

EXPERIMENT 73.

HELIOTROPIC REACTION OF PLANT WITH GRAVITY NEUTRALIZED.

If the influence of gravity is removed from a plant as in Experiment 67, and the light allowed to fall parallel to the axis of the plant, the roots and shoots will be seen to take opposite directions in a plane parallel to its rays.

EXPERIMENT 74.

THERMOTROPISM.

Grow seedlings of Corn in a pot, and place in a position where the light will be received perpendicularly. At a distance of 40 cm. place a sheet of smoked tin which is kept warm by a spirit-lamp. In twenty-four hours the shoots will have inclined toward the source of the heat. Repeat the experiment with Peas.

IRRITABILITY.

47. Periodic Movements.—The sun is continually changing its position during the day; consequently if a leaf remains in a

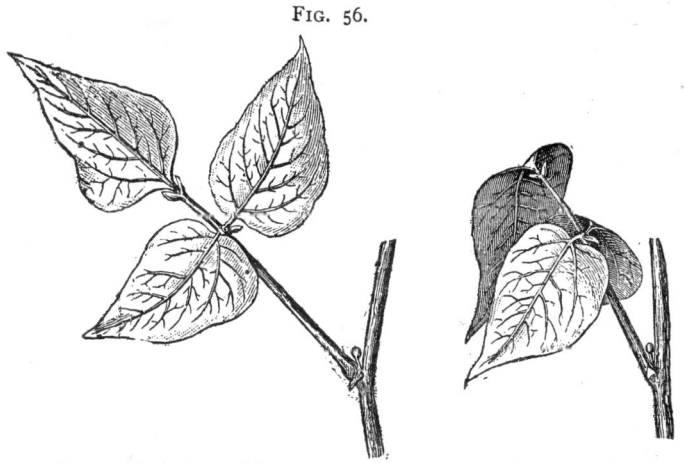

FIG. 56.

Day and night positions of leaflets of Bean. (Detmer.)

fixed position it receives the maximum heat and light at one moment only. It is found that leaves not only exhibit movements corresponding to the heat and light received, but also assumed certain positions to avoid excess or loss of heat. An organ loses or receives the least heat when its long axis is vertical. A great number of these movements have been described as "sleep movements."

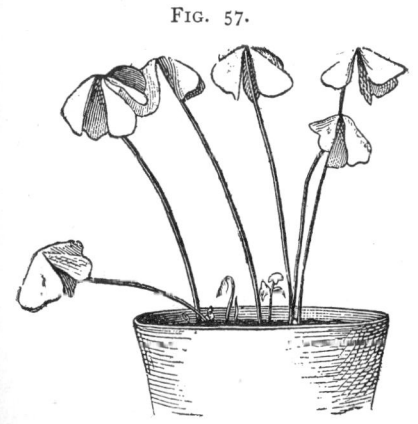

FIG. 57.

Sleep position of leaves of Oxalis induced by artificial darkness. (Hansen.)

EXPERIMENT 75.
SLEEP MOVEMENTS.

Observe the positions of the leaflets of a seedling of Bean or of an Oxalis, growing in the sunlight, at 8 A.M., 1 P.M., and 6 P.M. Determine whether these positions are due to light or heat by use of the dark chamber. (Figs. 56 and 57.)

48. Hydrotropism.—The moisture of the medium which surrounds the plant induces movements in certain organs either toward or away from the source of the moisture. The property of an organ by which it reacts to moisture is termed *hydrotropism.* By this power roots direct their apices toward portions of the soil containing the proportion of moisture best suited to their specific needs.

EXPERIMENT 76.
HYDROTROPISM OF ROOTS.

Cover a zinc box, 5 cm. wide and 20 cm. long, open on two sides, with gauze after it has been filled with moist sawdust, containing swollen seeds of Bean, Pea, or Corn. Suspend the apparatus under a pasteboard box, so that it hangs at an angle of 45 degrees. After a time the roots issue through the openings in the gauze beneath ; they do not follow geotropism and grow directly downward, however, but press against the layer of moist sawdust. Place the apparatus in a damp chamber where the moisture is equal in all directions from the roots, and they grow directly downward in response to the stimulus of gravity. In this case the roots receive the same stimulus from moisture in all directions, and in consequence no reaction to it is shown. They are free to respond to their progeotropic tendency.

FIG. 58.

Hydrotropism of roots. (Detmer.)

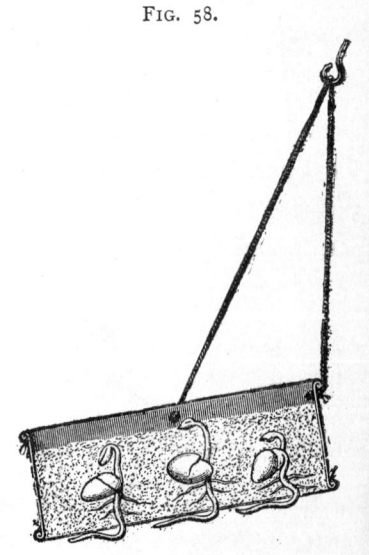

49. Contact Movements.—Many plants will exhibit movements so rapid as to be visible to the naked eye when touched or struck with any hard object. These movements serve various purposes in different groups of plants. In some instances, as in the Mimosa, this is a device for protecting the leaves

IRRITABILITY.

from injury. By this "sensitiveness" of tendrils, climbing plants are able to attach themselves to supports and lift their leaves to sunlight. In certain carnivorous plants, such as Drosera and Dionæa, the rapid movement of the tentacles and leaves enables these plants to capture insects which are held and whose substance is absorbed by the plant.

EXPERIMENT 77.
MOVEMENTS OF SENSITIVE-PLANTS.

Grow Mimosa pudica (Sensitive-plant) from seed, in a pot. Moisture and temperature of about 20° C. are necessary for the welfare

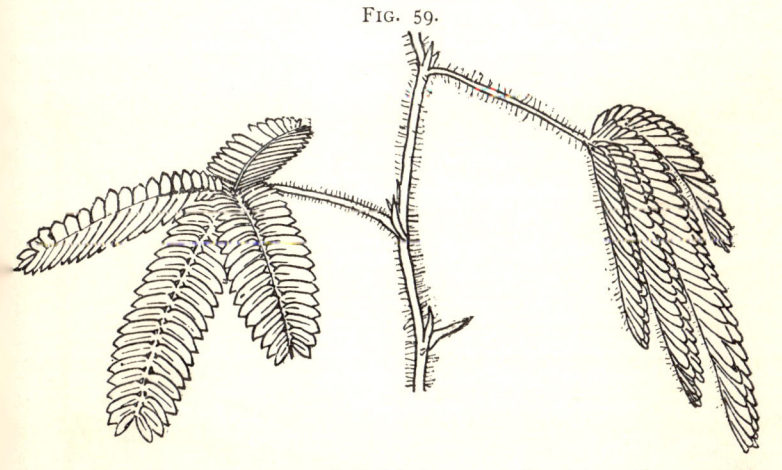

FIG. 59.

Mimosa pudica. The leaf on the left is in a normal position; the one on the right has been stimulated. (Detmer.)

of the plant; consequently it should be kept under a bell-jar slightly raised at one side to allow for ventilation, and placed in the sunshine. Try the following experiments: *a.* Jar the entire plant by striking the pot. In a few seconds the leaves take the position shown in Fig. 59. *b.* Strike one of the terminal leaflets. The pairs of leaflets fold up together in succession, and finally the whole leaf sinks on its petiole. *c.* Touch the upper side of the pulvinus with a pointed object. No movement follows. *d.* Touch the under side in like manner. A movement results. *e.* With a sharp knife cut off the petiole just above the pulvinus. A drop of water issues from the lower surface, which in an uninjured leaf would pass into the leaf-stalk.

64   *EXPERIMENTAL PLANT PHYSIOLOGY.*

EXPERIMENT 78.

MOVEMENTS OF STAMENS.

Touch stamen-filaments of Centaurea, Carduus, or Cichorium. The filaments contract.

FIG. 60.

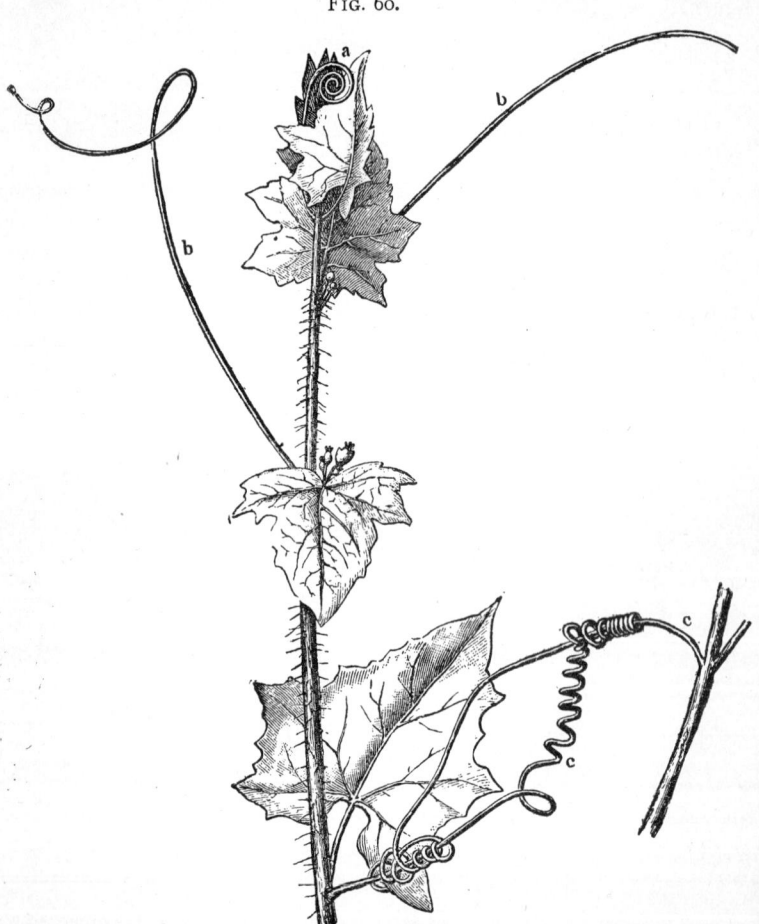

Tendrils of Bryony. (Kerner.) *a*, young tendrils; *b b*, nearly mature and very highly irritable; *c c*, two tendrils which have intertwined.

EXPERIMENT 79.

CURVATURE OF TENDRILS.

Touch a tendril of the Passion-flower, Bryony, or Squash on the concave surface near the tip with a pencil and observe carefully.

IRRITABILITY.

In a time varying from 30 seconds to several minutes a curvature is begun. Note rapidity, extent, and duration. Place a small rod in contact with the tendril. In a few hours it will have coiled around it. Observe the formation of spirals in the free portion of the tendril. (Fig. 60.)

EXPERIMENT 80.

RELATION OF HARDNESS OF OBJECTS TO CURVATURE PRODUCED IN TENDRILS.

Test the effect of water, mercury, soft gelatine, glass, iron, and wooden objects when brought in contact with tendrils.

EXPERIMENT 81.

ACTION OF LEAVES AND TENTACLES OF CARNIVOROUS PLANTS.

Obtain several plants of Sundew (Drosera) from the swamps. In digging them, care should be taken to leave a large mass of the soil on the roots of each so that their growth may not be greatly disturbed. Cover with a bell-jar and place in the sunlight. Touch

FIG. 61.

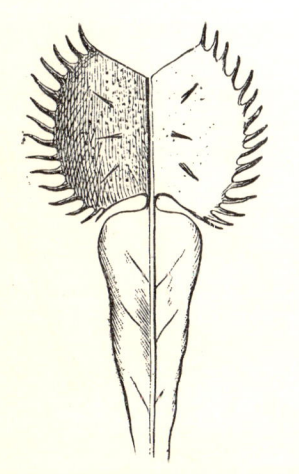

FIG. 62.

Leaf of Dionæa expanded. (Kerner.)

Leaf of Drosera with right leaf half contracted. (Darwin.)

the tentacles with small pieces of a large variety of substances, wood, sugar, starch, paste, alkali, meat, bits of stone, etc., and note to what substances the tentacles react and the rapidity of movement. Repeat with Dionæa.

50. Circumnutation.—If the tip of a shoot of some rapidly-growing plant, such as the Pea or Bean, is kept under observation for several minutes, it will be seen that it slowly changes position; and if the time of observation is extended, it will be found that it inclines successively toward every point in the horizon. In some plants the movement is in the same direction as the hands of a watch, and in others it is in the contrary direction. This nutatory movement of growing tips is quite generally distributed among plants, but it is most marked in twining stems. The causes which produce the movement are chiefly inequalities in growth extension of the sides of the stem, and the reaction to the influence of gravity.

EXPERIMENT 82.

CIRCUMNUTATION OF SHOOTS AND TENDRILS.

Note the positions of a growing tendril of the Gourd, Pea, Bryony, or Wild Balsam Apple at intervals for three hours.

Plant three or four seedlings of the Scarlet Runner or common Bean at equal distances from one another in a circle around an upright post. Mark the successive positions of the tips of each until it becomes twined around the support.

51. Hygroscopic Movements.—Many plants are provided with cells which take up or lose water in such manner as to give rise to very marked movements in the organs of which they form a part. Such cells are found in the leaf-blades of a large number of Grasses, and other plants which inhabit arid regions. In such plants this is a provision for rolling up the leaves in a form which will prevent undue loss of moisture from the organ. By a similar action many anthers open and allow the escape of the pollen, and fruit-capsules allow seeds to escape. In the latter instance sufficient force is sometimes furnished to throw the seeds to a distance or bury them in the soil.

IRRITABILITY.

EXPERIMENT 83.

WARPING OF WOOD.

Wipe the adhering moisture from a thin piece of wood, such as a cigar-box lid, which has lain in water 24 hours, and fasten to another piece of similar size which is air-dry, by means of a number of small nails. Lay in a dry place. The loss of moisture from the saturated piece will cause the double board to become curved.

Note the "warping" of unseasoned timbers.

EXPERIMENT 84.

MOVEMENTS OF FRUIT-CAPSULE OF IMPATIENS.

Bring some fruit of Balsamina (Impatiens noli tangere) which is nearly ripe into warm dry air. (Hold at a distance above a gas-flame.) The outer covering of the fruit contracts and forcibly ejects the fruit. Make sections of the portions of the capsule, and describe the action of the hygroscopic cells.

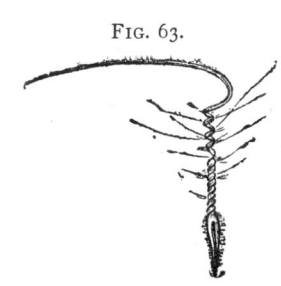

FIG. 63.

Erodium seed with the long beak twisted by drying. (Detmer.)

EXPERIMENT 85.

TWISTING MOVEMENTS OF THE BEAK OF AN ERODIUM SEED.

A moist seed of Erodium is placed with the point in a damp soil. In drying the beak curves and twists in a spiral form. If the twisting of the beak is hindered by a piece of wood thrust into the sand beside it, the force will be exerted upon the seed, and will be thrust into the sand still deeper. (Fig. 63.)

CHAPTER VI.

GROWTH.

52. Nature of Growth.—The increase of the living substance of an organism is designated *growth*, and it is generally accompanied by an increase in weight and size. This increase does not, however, always accompany growth; indeed, it was demonstrated in Experiment 29 that a plant may grow while losing in weight. If the plant is accumulating storage material it will, on the other hand, undergo an increase in weight not in any manner connected with growth. It is also to be noted that independent changes in form and size occur which are due simply to alterations in the force of turgor and in the extensibility of the cell-walls. Lastly, growth does not consist in the formation of new cells; on the contrary, the formation of new cells is a result of growth.

EXPERIMENT 86.

MEASUREMENT OF GROWTH EXTENSION.

To determine the increase in length of a plant the simple auxanometer shown in Fig. 64 will be found fairly accurate. This apparatus consists of an upright stand 50 cm. in height to which is attached a horizontal arm 10 cm. in length. To the end of this arm is attached a wooden pulley 4 cm. in diameter, in such manner that it will turn freely. To one side of this pulley is fixed a thin wooden pointer 20 cm. in length. This pointer is made from a strip not more than 2 mm. in thickness, and has wrapped around the larger end at g a sufficient quantity of tinfoil to balance the longer end. A curved paper-scale ruled to 2 mm. is held by another stand near the tip of the pointer. A linen or silk thread is tied to the tip of a shoot of a Coleus, Tomato, or Potato, or leaf of a Narcissus grown in a pot. The plant is set directly under the pulley and the string is passed over the pulley, and attached to this end is a weight G of one gram or more to keep the threat taut. As the plant grows in length it allows the weight G to descend, turning the pulley as it does so. The pointer is attached directly to the pulley, and as the elongation takes place it

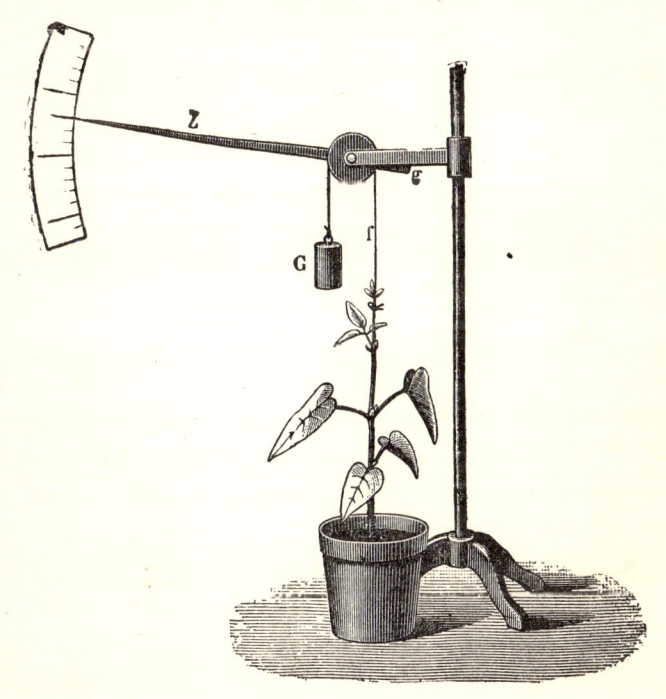

FIG. 64.

Lever auxanometer. (After Oels.) Z, lever; g, balance-weight on lever; G, counterpoise to keep the string taut; f, string.

passes downward along the scale. Since the length of the pointer from the center of the pulley is 16 cm. and the radius of the pulley is 2 cm., the amount of growth is magnified 8 times. The apparatus is set up with the pointer at zero on the upper end of the scale. Observations of its position should be made at least three times daily. A growth extension of 1 to 5 cm. daily may be expected under favorable circumstances.

EXPERIMENT 87.

MEASUREMENT OF ROOT EXTENSION.

Germinate Pea, Bean, or Squash seeds until the primary roots are 2 cm. in length. Place one of the seedlings in the bowl of a thistle-tube or small funnel, with the root depending downward in the tube. Cover the seedling with moist cotton and place the bottom of the tube, in a vessel of water. By means of India ink mark off intervals of 2 mm. on the tube, and set the whole apparatus in equal-sided light so far as possible. Note the position of the root-tip at least twice daily. A growth of 4 to 20 mm. in a day may be expected.

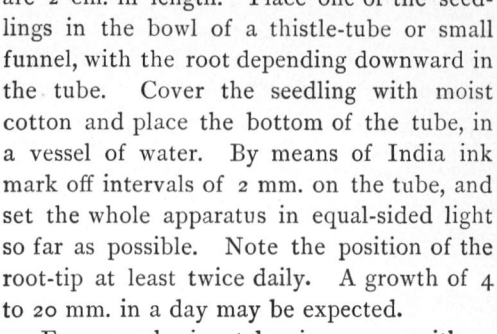

Focus a horizontal microscope with a power of 25 diameters on the extreme tip of the root. It will be seen to move slowly across the field of view. (Fig. 65.)

EXPERIMENT 88.

MEASUREMENT OF GROWTH INCREASE BY WEIGHT.

Select a young Squash or Pumpkin which has attained a diameter of a few centimeters and place on the pan of a druggists' balance (Fig. 24), with the vine supported in such a manner that it bears as little weight as possible on the balance. Place in the second pan sufficient weights to establish an equilibrium. Equalize the scale morning, noon, and evening, and the amount of increase may be directly obtained. During the period of most rapid growth the daily increase will amount to 200 to 700 grams. At times the weight of the fruits will be found less at noon than in the morning owing to excessive evaporation of water from its surface and that of the leaves.

FIG. 65.—Seedling of Squash in a thistle-tube. (Detmer.)

53. Grand Period of Growth.—With regard to growth three regions may be observed in any organ composed of many cells: one in which new cells are constantly forming, as in the tips of roots and shoots; another in which the cells are increasing in size, and a third in which the cells have attained their full size and maturity. The time inclusive of the formation and enlargement of a cell is termed its *grand period of growth.* In the case of an organ, this period includes the times from the formation of all of its cells to their maturity. All of the cells are not formed at the same time and do not reach maturity at the same time. The portion containing the cells which are enlarging most actively is designated the *zone of maximum growth.* This zone is constantly changing its position, as may be seen in the following experiments.

EXPERIMENT 89.

ZONE OF MAXIMUM GROWTH OF ROOTS.

Select a healthy seedling of Pea, Bean, or Squash with a rootlet 2 cm. in length. With a pointed camel's-hair brush mark off ten intervals 1 mm. apart. Place the seedling in a thistle-tube as in Experiment 87. Set where it may receive an equal-sided illumination. In twenty-four hours observe the length of the intervals. The fourth, fifth, sixth, and seventh from the tip will be found to have elongated much more than any of the others. Twenty-four hours later the terminal division will have partaken of this elongation, showing that the zone of maximum growth moves steadily toward the tip. Now follow the growth of the second interval. At the beginning of the experiment it elongates somewhat slowly at first, then more rapidly, until it is growing more rapidly than any other portion of the root. Its rate then decreases until it finally ceases. In the mean time the next interval toward the tip begins to increase in rapidity,

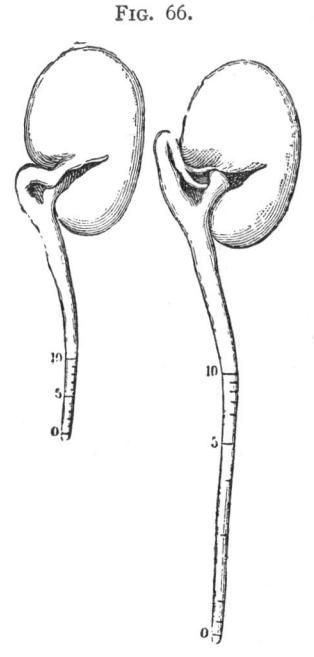

FIG. 66.

Seedlings of Pea. (Sachs.) Showing zone of maximum growth.

and about the time the previous one has begun to lessen its rapidity of growth, it has reached its maximum. In this manner the zone of maximum growth progresses. (Fig. 66.)

EXPERIMENT 90.

ZONE OF MAXIMUM GROWTH OF STEMS.

Growth of stems may be observed in the same manner as in the last experiment. The elongation of the natural divisions, the *internodes*, can be measured and compared with one another. The elongating part is greater than in roots—35 millimeters in the Bean. The internodes vary in length; the middle ones are the longest. Satisfactory results may be attained by the measurement of centimeter intervals on the stem of Bean or Corn. Compare movement of zone of maximum growth with that in roots.

EXPERIMENT 91.

ZONE OF MAXIMUM GROWTH OF LEAVES.

Cultivate Gourd or Tobacco plants in large pots, and after some leaves have been formed place them under large bell-jars, and set in light, but not in direct sun light, in a temperature as nearly constant as possible. Before doing this mark off on the petiole or midrib of a young leaf a scale as above. Compare observations with results of above experiments.

54. Influence of Light on Growth.—While light is necessary for the formation of food by photosynthesis, and for the performance of certain other functions, it at the same time generally retards growth. Only the blue-violet end of the spectrum exercises this retarding influence. By reason of this influence the maximum growth of a great number of plants occurs after they have been deprived of light for the longest period, which is in the morning, or just before daylight. Temperature is generally more favorable to growth during the afternoon, and as a consequence the plant grows rapidly at this time also. In fact the maximum growth often occurs then.

GROWTH.

Light also influences the form and size of the cells, as well as of the entire plant. (See § 40.)

EXPERIMENT 92.

GROWTH OF SEEDLINGS IN DARKNESS.

Grow seedlings of Cucurbita (Squash) in similar pots, some of which are set in the light, and others are covered by a pasteboard box, at the same temperature. The latter do not develop normally; the shoot axes are much extended, and form only imperfect leaves, which assume an upright position. (Fig. 67.) All parts of the plant are pale and disproportionately tender. The lignification of the wood is hindered, in consequence of which there is no opposition to the extension of the tissues by the turgor stretching of the parenchyma-cells.

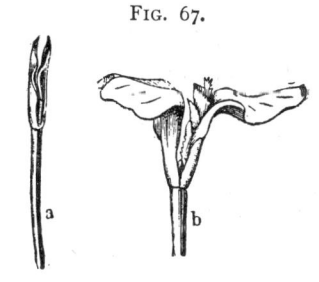

FIG. 67.

Cucurbita seedlings. (Detmer.) *a*, grown in darkness; *b*, grown in light.

Remark.—Care must be taken in this experiment that both plants do not stand in the sunlight, otherwise an abnormally high temperature will arise in the pasteboard box, and thus the relations of temperature will be altered.

EXPERIMENT 93.

COMPARISON OF GROWTH OF SEEDLINGS IN LIGHT AND DARKNESS.

Germinate a number of Peas in a pan of moist sawdust until the main roots are 2 cm. long. After the roots of several, as nearly alike as possible, have been marked with a scale, as in Experiment 89, place some in the light, and others under a pasteboard box, over spring-water. It will be found that the growth in light is less than in darkness. At the same time the daily period of growth can be observed.

55. Influence of Light upon the Anatomy of the Leaf. The leaves of common trees have in the upper side a closely-arranged layer of cells rich in chlorophyll (*palisade parenchyma*), and in the lower side a loosely-arranged tissue poor in chlorophyll (*spongy parenchyma*). This arrangement depends upon the influence of light. Shaded leaves exhibit another structure, and leaves which have been artificially twisted, so that the lower side is exposed to the light, reverse the arrangement of these two kinds of parenchyma.

EXPERIMENT 94.

INFLUENCE OF LIGHT ON THE STRUCTURE OF LEAVES.

Turn and fasten young leaves of the Beech (Fagus sylvatica) so that the under side is exposed to the light. When mature they show palisade tissue in the side now above, and spongy parenchyma in the side turned away from the light, as may be seen on examination with the microscope.

EXPERIMENT 95.

DEVELOPMENT OF FLOWERS IN DARKNESS.

Enclose a young inflorescence of Scarlet Runner or Morning-glory in a pasteboard box or bag of thick black cloth. The flowers and fruit will develop normally in the darkness thus secured.

56. Influence of Gravity and Light on the Formation of Organs.—Light and gravity influence the origin and demarkation of the forms of organs in a very remarkable manner. If a twig of Willow or some other plant is placed in a damp chamber, root and leaf buds will develop under the bark. If the twig is placed in an upright position, the roots will develop below and the leaves above. This "polarity" is, according to Vöchting, due to light and gravity. The action of light induces the formation of shoots on the illuminated side, and roots on the shaded portion. That gravity acts in a similar manner may be shown under other conditions. If a Willow twig is rapidly turned, like the diameter of a wheel (Experiment 68), shoots will be formed near the center of revolution, and roots at the peripheral ends. The *symmetry* of flowers, according to Vöchting's researches, is due to the influence of gravity.

EXPERIMENT 96.

COMPARISON OF THE GROWTH OF CUTTINGS IN LIGHT AND DARKNESS.

Fasten a Willow twig in an upright position in a covered glass cylinder containing some water, and set in the sunlight. It develops roots below and shoots above. Another twig treated in the same manner but placed in the dark acts similarly. (Fig. 68.)

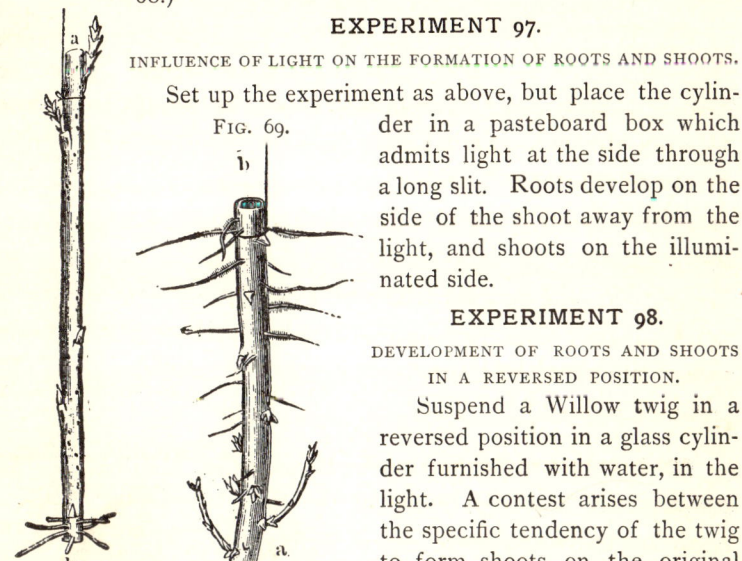

EXPERIMENT 97.

INFLUENCE OF LIGHT ON THE FORMATION OF ROOTS AND SHOOTS.

Set up the experiment as above, but place the cylinder in a pasteboard box which admits light at the side through a long slit. Roots develop on the side of the shoot away from the light, and shoots on the illuminated side.

EXPERIMENT 98.

DEVELOPMENT OF ROOTS AND SHOOTS IN A REVERSED POSITION.

Suspend a Willow twig in a reversed position in a glass cylinder furnished with water, in the light. A contest arises between the specific tendency of the twig to form shoots on the original upper end, and roots on the lower end, and the influence of light and gravity which directly oppose it. At first the habit of the plant prevails, and roots are formed on the upper, shoots on the lower end. Then the influence of the physical forces is manifested by the development of roots on the lower and shoots on the upper end of the twig. (Fig. 69.)

68. Willow twig in normal position. (Hansen.) *a*, shoots ; *b*, roots.

69. Willow twig in reversed position. (Hansen.) *a*, original upper end ; *b*, original lower end.

57. Influence of Temperature on Growth.—For every plant there are five important temperature divisions : 1st, destructive cold, a low temperature producing death by the disorganization of the protoplasm ; 2d, *specific zero*, which arrests the activity of the protoplasm, but does not necessarily result in damage to the organism; 3d, *optimum temperature*, in which normal development proceeds; 4th, maximum temperature, at which the protoplasmic activity comes to a standstill without necessarily injuring the organism; 5th, destructive heat, producing death by disintegration of the protoplasm. These divisions vary greatly with each species.

58. Sources of Heat.—The temperature of any plant is the result of the heat it receives from several sources. A portion comes directly from the heat-rays of the sunlight, as well as from the light-rays which it is able to convert into heat by means of chlorophyll, anthocyanin, and other coloring matters. Another portion is received from the soil, which is generally more constant in temperature than the air. According to Kerner the soil of a mountain at a height of 2200 meters is 3.6° C. higher than the surrounding air. Another and by no means unimportant source of heat is the combustion of the carbon compounds in the plant. (See Experiment 53.)

59. Influence of Temperature on Geographical Distribution.—In consequence of the obliquity of the ecliptic, no place on the earth has the same temperature during the entire year, disregarding even the changes of day and night. Fluctuations in temperature vary greatly with the locality: it is greater in valleys and at the poles than it is on mountains and at the equator. Fluctuation further depends upon the continental or oceanic position of a place. Again, between the elevated cold regions of the warmer zones and the polar regions there is the difference of short period of daylight and the long summer on one hand and the longest period of daylight, and a short summer on the other. These conditions of temperature, together with those of rainfall and soil, are the most important factors in the geographical distribution of plants. The regions which are not subject to extremes of temperature will be found most suitable for the greater number of species. While some species of plants thrive with a low summer heat if the temperature does not sink to the destructive point in winter, others endure a low temperature in winter very well if the temperature ascends high enough in summer to permit normal fruit-formation.

60. Freezing of Plants.—Formerly it was believed that the cell-sap was frozen by cold, that by the resultant expansion the cell-walls were torn, and in this way the plant was killed. It has, however, been demonstrated that a mechanical destruction of the cell by rupture does not take place, for the ice-formation goes on only in the intercellular spaces, or, in the simpler plants, in the water thrown out around the plant. It is therefore now held that death by cold is the result of a chemical process, which can occur at a temperature even above freezing-point.

Rapid or slow thawing of frozen plants has no influence upon the life-energy of the plants. If, however, frozen plants which are not killed are thawed slowly, the cells can reabsorb the water from the melting ice-crystals around them and regain their former turgor. If thawed rapidly, a portion of the water of the ice-crystals is evaporated or driven away, and the cells cannot regain their turgor. When a plant remains frozen for some time, the water slowly evaporates from the crystals and the plant is eventually dried. Therefore frozen plants may be killed by loss of water, either through continued cold or rapid thawing.

Salt solutions freeze at lower temperatures than pure water, and in their freezing the water is separated out in the form of crystals. Cell-sap, a solution of several stable substances in water, acts similarly, and plants may therefore endure a temperature many degrees below freezing-point without being frozen.

EXPERIMENT 98.

FREEZING OF A SALT SOLUTION.

Partially freeze a solution of potassium bichromate or copper sulphate. The frozen portions are distinguished from the concentrated fluid by the paler color. The freezing begins at a temperature a few degrees below zero C.

EXPERIMENT 99.

FREEZING OF A BEET.

Place a section of a Beet, a centimeter thick, well washed and dried, in a dish covered with a glass plate to prevent evaporation, at a temperature of 6 degrees below zero C. When the section is frozen, the surface will be covered with a layer of ice, which when examined with the microscope, at a temperature below zero, will be found to consist of parallel crystals. A very heavy ice-layer is found on the under side of the section, where it has been in contact with the dish. The ice is not colored red, proving that not cell-sap but pure water drawn from the cell has been frozen. No rupture of the cell-wall occurs in the freezing of living cells, as would be the case if the enclosed fluid were frozen.

EXPERIMENT 100.

FREEZING OF SPIROGYRA.

Freeze some Spirogyra filaments in a drop of water on a glass slide. After thawing no rupture of the cell-walls appears.

EXPERIMENT 101.

FREEZING OF POTATOES.

Place some Potatoes in a temperature of 5 to 10 degrees below zero centigrade, over night. They are frozen hard, and upon thawing become very soft, allowing the sap to be forced out by the lightest pressure. Their power of germination is lost, and they easily rot. Whether the Potatoes are thawed quickly or slowly is a matter of indifference.

61. Relation of Moisture to Freezing.—Low and high temperatures are destructive to plants and plant-organs in proportion to their richness in water.

EXPERIMENT 102.

FREEZING OF PEA, BEAN, AND WHEAT.

Place some air-dry seeds of the Pea, Bean, or Wheat for several hours in a temperature of 5 to 10 degrees below zero centigrade. They do not lose the power of germination, as may be shown. The same kinds of seeds when saturated with water are killed by this temperature, and are unable to germinate.

Remark.—Trees behave similarly. In winter, when they contain but little water, they endure a high degree of cold; a late spring frost kills them, because the trunks and twigs are full of sap.

EXPERIMENT 103.

EFFECT OF HIGH TEMPERATURE ON SATURATED SEEDS.

Place 30 swollen seeds of Peas or Wheat for a quarter of an hour in water at a temperature of $60°$ to $70°$ C., and then place in a germinator. They do not germinate, while 30 other seeds placed in the germinator after soaking develop normally.

62. Protoplasm which has been killed by low or high temperature undergoes molecular changes; it then becomes permeable to acids and coloring matters. (See § 60.)

EXPERIMENT 104.

ESCAPE OF CELL-SAP OF BEET KILLED BY LOW TEMPERATURE.

Frozen and unfrozen pieces of Beet are placed in water; the first colors the water red, the latter does not. The protoplasm of the frozen cells allows the colored sap to pass through it.

EXPERIMENT 105.

ESCAPE OF SAP FROM A BEET KILLED BY HIGH TEMPERATURE.

Perform the above experiment, using pieces of Beet, one of which has been in water at a temperature of $60°$ to $70°$ C. The result is the same as in Experiment 104.

EXPERIMENT 106.

ESCAPE OF CELL-SAP CONTAINING OXALIC ACID FROM A STEM OF BEGONIA KILLED BY HIGH TEMPERATURE.

Place two pieces of a petiole of Begonia in distilled water after one of them has been treated with water at a temperature of $60°$ to $70°$ C. until colorless. Add a solution of calcium chloride to the dishes containing the pieces. The water in one dish remains clear, while that in the other becomes turbid from the formation of oxalate of calcium. The heated portion permits the escape of oxalic acid which it contains.

63. Loss of Heat.—On account of the importance of warmth for the chemical processes in the building up of the plant, many plants possess peculiar adaptations for preventing undue loss of heat.

EXPERIMENT 107.
ADAPTATIONS TO PREVENT LOSS OF HEAT.

Grow seedlings of Helianthus (Sunflower) and Cucurbita (Squash). As soon as the cotyledons are raised above the earth, it may be observed that they are extended during the daytime, and during the coolness of the evening close together above, whereby the loss of heat by radiation is decreased. (See Experiment 75.)

64. Resting Period.—It is known that the winter buds of trees and shrubs can be made to open very early in the spring if they are placed in a warm room or greenhouse. In this way, shoots cut from Syringa vulgaris (Lilac), or the Willow, in February, may be given an early development. It might be inferred that these plants are compelled to rest by the winter cold and need only heat to set in motion their normal development. This is not, however, entirely true. Experiments have shown that the winter resting period is necessary for the plant, or rather that it has become accustomed to it by thousands of years of habit. It is on account of this acquired habit that buds brought into a warm room in January do not begin to develop before March, and Potato-tubers brought into a warm room in the autumn do not begin to germinate until after a resting period of greater or less duration. Potato-tubers which are placed in a temperature of zero centigrade, for four weeks immediately after digging in August, upon being planted in garden soil and watered, will begin the development of buds.

GROWTH.

EXPERIMENT 108.
ACCELERATED DEVELOPMENT OF SHOOTS.

Cut off twigs of Syringa (Lilac), Cornus (Dogwood), Salix (Willow), etc., in several winter months ending with February, and place them in water in a warm room, or, better, under a bell-jar to keep them moist. The development of the buds proceeds according to the laws given above.

65. Correlation Processes.—Not all the shoots of a plant come to full development. Only the strongest and most useful to the whole plant develop, while the others either perish or carry on a kind of "sleep-life." These last are generally styled *latent buds*. If the plant is robbed of a "concurrent" organ by any accident, the nourishment heretofore used by that organ is sent to a latent bud, which then emerges from its period of rest and develops. Such phenomena are termed *correlation processes*. In gardening much use is made of this capacity of the plant; as, for example, in the formation of thick hedges, in the development of branches and flowers on the Fuchsia, etc.

FIG. 70.

EXPERIMENT 109.
DEVELOPMENT OF LATERAL SHOOTS OF THE BEAN.

Germinate two plants of Bean in pots. Cut the epicotyl from one as soon as it appears above the ground. Then the buds in the axils of the cotyledons develop instead. The other plant serves as a means of comparison.

EXPERIMENT 110.
DEVELOPMENT OF LATERAL BUDS OF THE POTATO.

Place a Potato-tuber with the stem-scar underneath in a warm room without the addition of water. The buds near the top develop. Cut these off and the lower ones start into active growth. (Fig. 70.)

Sprouting Potato. (Detmer.)

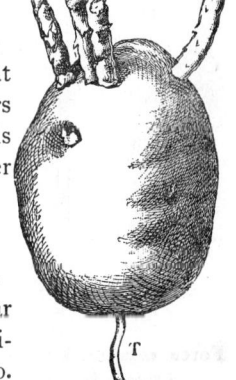

66. External Mechanical Force Exerted by Growing Organs.—The growing cells of plants are able to exert a pressure on bodies surrounding them which may amount to from 12 to 15 atmospheres. By this force roots and other fixing and absorbent organs are driven through the soil, and aerial organs push their way upward through the air. The total amount of energy used in the performance of external work during the lifetime of the plant is very great. The spore-bearing cap of a Mushroom has been known to lift a weight of 160 kilograms. A root of Larch 30 cm. in diameter has lifted a stone 1600 kilograms in weight, while a root of a germinating Bean has exerted a lateral pressure on the soil amounting to 1.5–4 kilograms. All growing organs expand with similar force, but in the examples given the form of the organ is such as to utilize the force in penetrating the substratum. The growing fruit of a Cucurbita is capable of exerting a pressure of several thousand kilograms, though it ordinarily meets with no resistance.

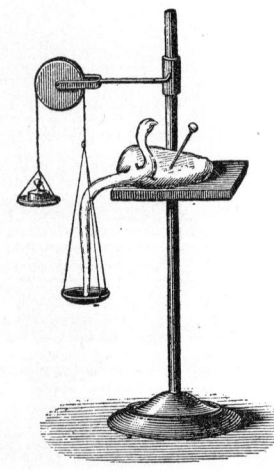

FIG. 71.

Force exerted by growing roots. (Mangin.)

EXPERIMENT 111.

POWER OF PENETRATION OF RHIZOIDS OF A HEPATIC.

If a Hepatic is placed on several folds of moist filter-paper in a chamber saturated with moisture, within forty-eight hours the rhizoids will have pierced the filter-paper. The holes through which the rhizoids have penetrated were certainly not there before. The fibrous structure of the paper is so dense that a starch-grain of corn, which is only two micromillimeters in diameter, cannot find its way through, yet the rhizoids, which are 10 to 35 micromillimeters in diameter, easily accomplish it.

EXPERIMENT 112.

FORCE EXERTED BY GROWING ROOTS.

To a small upright stand attach a horizontal arm bearing a small wooden pulley. Fasten a scale-pan to a cord passing over the pulley to the other end of which is attached a second pan containing a 5-gram weight. Fill the first pan firmly with moist sand, and fasten a seedling of Bean in such position that it touches the sand. It will push downward into the sand and elevate the weight-pan. The scale-pan touched by the root may be suspended directly from the horizontal arm by a delicate spiral spring, omitting the pulley and second scale-pan. As the root grows it will push the pan downward as before, and the distance through which the scale-pan moves will indicate the force directly. The strength of the spring can be determined by placing weights on the scale-pan. (Fig. 71.)

APPENDIX.

ENGLISH AND METRIC WEIGHTS AND MEASURES.

LENGTH.

1 micro-millimeter = $\frac{1}{1000}$ millimeter or $\frac{1}{25000}$ inch.
1 millimeter (mm.) = $\frac{1}{25}$ inch.
1 centimeter (cm.) = 10 mm. = $\frac{2}{5}$ inch.
1 decimeter (dm.) = 100 mm. = 4 inches.
1 meter = 1000 mm. = $39\frac{1}{3}$ inches.
1 inch = 25 mm.
1 foot = 305 mm. or $30\frac{1}{2}$ cm.
1 yard = .91 meter.

WEIGHT.

1 gram = $15\frac{1}{2}$ grains.
1 kilogram = 1000 grams = 32 oz. Troy or $35\frac{1}{4}$ oz. Avoirdupois.
1 oz. Troy = 31 grams.
1 oz. Avoirdupois = 28 grams.
1 lb. Avoirdupois = 450 grams.

CAPACITY AND WEIGHT.

1 gram = 1 cubic centimeter (cc.) = $15\frac{1}{2}$ grains.
1 liter = 1000 grams, or 1000 cc., or 1 kilogram = $35\frac{1}{4}$ oz. Avoirdupois or 32 oz. Troy.
1 pint = 20 oz. Avoirdupois = $567\frac{1}{2}$ grams or $567\frac{1}{2}$ cc.

CAPACITY (VOLUME).

1 liter = 1000 cc. = 1 cubic dm. = $1\frac{3}{5}$ pints.
1 pint = 36 cubic inches = $567\frac{1}{2}$ cc.
1 gallon = 8 pints = $4\frac{1}{2}$ liters.
1 cubic foot = 6 gallons = $28\frac{1}{3}$ liters.

APPENDIX.

CENTIGRADE AND FAHRENHEIT THERMOMETER SCALES.

− 30° Cent. = − 22° Fahr.	40° Cent. = 104° Fahr.
− 25° Cent. = − 13° Fahr.	45° Cent. = 113° Fahr.
− 20° Cent. = − 4° Fahr.	50° Cent. = 122° Fahr.
− 15° Cent. = 5° Fahr.	55° Cent. = 131° Fahr.
− 10° Cent. = 14° Fahr.	60° Cent. = 140° Fahr.
− 5° Cent. = 23° Fahr.	65° Cent. = 149° Fahr.
0° Cent. = 32° Fahr.	70° Cent. = 158° Fahr.
5° Cent. = 41° Fahr.	75° Cent. = 167° Fahr.
10° Cent. = 50° Fahr.	80° Cent. = 176° Fahr.
15° Cent. = 59° Fahr.	85° Cent. = 185° Fahr.
20° Cent. = 68° Fahr.	90° Cent. = 194° Fahr.
25° Cent. = 77° Fahr.	95° Cent. = 203° Fahr.
30° Cent. = 86° Fahr.	100° Cent. = 212° Fahr.
35° Cent. = 95° Fahr.	110° Cent. = 230° Fahr.

INDEX TO PLANT NAMES.

	PAGE
Algæ	6
Anthemis	45
Apple	26
Aroids	45
Asclepias	32
Bacteria	3, 11, 12, 44
Balsamina	67
Barley	48
Bean,	5, 10, 27, 41, 52, 53, 59, 60, 62, 66, 70, 72, 81, 83
Beech	14, 74
Beechdrops	11
Beet	78, 79
Begonia	25, 49, 79
Bellis	45
Birch	31
Bryony	64, 66
Buckwheat	5
Cabbage	14
Carduus	64
Carrot	14
Centaurea	64
Cichorium	64
Coleus	14, 32, 69
Composite	45
Corallorhiza	11
Coral-root	11
Corn,	5, 8, 10, 20, 27, 34, 37, 52, 60, 62, 72
Cornus	81
Cucurbita	73, 80, 82
Cuscuta	10, 11
Dahlia	20
Dionæa	63, 65

	PAGE
Dodder	10
Dogwood	81
Drosera	63, 65
Elder	18, 33
Elodea	36, 37, 38, 40
Epiphegus	11
Erodium	67
Euphorbia	32
Fagus	74
Fritillaria	52
Fuchsia	81
Funaria	42
Fungus	11
Geranium	14, 20, 42
Gourd	66, 72
Grape	18, 19, 20, 36
Grass	53, 66
Helianthus	32, 59, 60, 80
Hemp	44
Hepatic	82
Hyacinth	52
Impatiens	11, 14, 34, 67
Indian-pipe	11
Iris	14, 24
Larch	82
Leontodon	45
Lilac	80, 81
Liverwort	14
Lonicera	29, 31, 32
Madia	45
Malva	59
Maple	59
Marchantia	25
Milkweed	32

87

INDEX TO PLANT NAMES.

	PAGE
Mimosa	50, 62, 63
Mistletoe	10, 11
Monotropa	11
Morning-glory	74
Mosses	14
Mould	11
Mushroom	11, 82
Mustard	8, 59
Myosotis	49
Narcissus	52, 69
Nettle	19
Oak	14
Onion	52
Oxalis	60
Passion-flower	65
Pea, 5, 8, 9, 10, 16, 17, 27, 45, 46, 52, 53, 55, 62, 66, 69, 70, 79	
Peony	49
Phaseolus	53, 59
Poplar	18
Potato	26, 69, 80, 81
Pumpkin	70
Raspberry	22
Rhubarb	18
Rose	49
Rosebush	21, 23
Rust	11

	PAGE
Salix	81
Sambucus	18, 31, 34
Scarlet runner	74
Seaweed	39
Sensitive plant	50, 63
Sinapis	59
Smut	11
Sonchus	32
Spirogyra	42, 78
Spurge	32
Squash	9, 64, 70, 73, 80
Sundew	65
Sunflower, 18, 19, 20, 29, 32, 60, 76, 80	
Symphytum	23
Syringa	80, 81
Toadstool	11
Tobacco	72
Tomato	14, 17, 41, 69
Touch-me-not	34
Tropæolum	41, 49
Tulip	52
Vaucheria	42
Wheat	5, 27, 46, 75
Wild balsam-apple	66
Wild lettuce	32
Willow	18, 34, 74, 75, 80, 81
Yeast	44